CW01596934

Introducing Systems Analysis and Design Volume 2

PUBLISHED BY NCC PUBLICATIONS

Keywords for information retrieval (drawn from the NCC *Thesaurus of Computing Terms*): computer projects, computer security, database, data control, file organisation methods, forms, design, systems analysis, systems design, systems documentation

British Library Cataloguing in Publication Data

Lee, Barry
Introducing systems analysis and design
Vol. 2
1. Systems analysis
I. Title
003 QA402
ISBN 0-85012-207-4

First published in 1979 by:
NCC Publications, The National Computing Centre Limited, Oxford Road, Manchester M1 7ED, England

Reprinted 1980
Reprinted 1981
Reprinted 1982

ISBN 0-85012-207-4

This book is set in 10/11pt. Times Roman Series
Printed in Great Britain by
Nene Litho, Earls Barton, Northamptonshire, and bound by Weatherby Woolnough, Wellingborough, Northamptonshire.

The National Computing Centre develops techniques and provides aids for the more effective use of computers. NCC is a non-profit distributing organisation backed by government and industry. The Centre

- co-operates with, and co-ordinates the work of, members and other organisations concerned with computers and their use
- provides information, advice and training
- supplies software packages
- promotes standards and codes of practice

Any interested company, organisation or individual can benefit from the work of the Centre by subscribing as a member. Throughout the country, facilities are provided for members to participate in working parties, study groups and discussions, and to influence NCC policy. A regular journal – 'NCC Interface' – keeps members informed of new developments and NCC activities. Special facilities are offered for courses, training material, publications and software packages.

For further details get in touch with the Centre at Oxford Road, Manchester M1 7ED (telephone 061-228 6333)

or at one of the following regional offices:

Belfast
1st Floor
117 Lisburn Road
BT9 7BP
Telephone: 0232 665997

Birmingham
7th Floor
Devonshire House
Great Charles Street
B3 2PL
Telephone: 021-236 6283

Bristol
6th Floor
41 Corn Street
BS1 1HG
Telephone: 0272 277077

Glasgow
2nd Floor
Anderston House
389 Argyle Street
G2 8LR
Telephone: 041-204 1101

London
11 New Fetter Lane
EC4A 1PU
Telephone: 01-353 4875

Contents

Part IV
Physical Design of Computer Subsystem

Once the logical system definition has been agreed by the users, the detailed design of the physical system can commence. This part describes the detailed physical design of the computer part of the information system. For ease of understanding, the computer subsystem design is broken down into five aspects: stored data, outputs, inputs, procedures, and security. In practice, because all these aspects are interrelated, they would not be designed in isolation, but in an integrated way. Similarly, though the next part of the book describes the detailed physical design of the manual subsystem separately, in practice the manual parts of the system would be designed in conjunction with the computer parts because integration is essential.

The first chapter of this part, Chapter 10, deals with computer file design and examines different types of file, file structure, file organisation and file access methods. Chapter 11 also deals with stored data but from the point of view of database design; thus, it discusses database objectives, the differentiation between logical and physical data structures, and methods of physical data organisation which are needed in database systems. Chapter 12 examines the design and specification of computer outputs and inputs. Chapter 13 is concerned with computer procedure design, including types of computer procedure, design tools and considerations, and the timing of computer procedures. The final chapter of this part is devoted to a discussion of system security from the viewpoint of the protection required and the techniques which are available to achieve security.

10 File Design

INTRODUCTION

The first requirement of physical design of the system is to decide how the logical data structures which have been defined in the logical system definition are to be physically stored on the backing storage devices. This chapter examines file design under a series of headings – types of file, file structure, file organisation, file access methods, and design. The next chapter, on 'database design', examines the design of integrated data storage in the form of interrelated records.

A file may be defined as a collection of items of data organised into records in such a way that specific items of data or records can be retrieved and accommodated in main storage when required for processing. A record is a group of related facts treated as a unit representing, perhaps, a particular transaction. An example of this is the payment of an invoice against goods received. The record of this transaction would probably contain the related items of the supplier's account number, invoice number, and the total amount paid, treated as a unit. A collection of similar records constitute a file, in this case a file of payment transactions. Each record has within it a *key* field which is used to identify the record. A glossary is appended for those readers who may require clarification of some technical terms used in the text.

TYPES OF FILE

It is useful to identify the various types of file.

Master files

Master files are the most important type of file; most file design activity focuses here. Master files contain essential system data which is retained permanently within the system and processed within each operational cycle. There are two categories of master files: reference and dynamic.

A *reference* master file consists of static records or records unlikely to change frequently: a customer's name, address and account number on a customer file, or product records comprising product description, code and price. These records are used for reference purposes, eg, to produce sales invoices, product catalogues, production orders, etc. A collection of all the customer records constitutes a customer file, and all the product records, a product file; such files are called reference master files.

This contrasts with a *dynamic* master file which contains records being continually changed (updated) by events or business transactions:

- a sales ledger file showing the sales of goods or services transactions, the receipts of money in payment and the current indebtedness of customers;
- a stock file showing the receipts, issues and current balance of each item of stock;
- a seat-reservation file showing the names of customers who have reserved seats and the current position of seats available.

The two types of master file may be kept as separate files, or they may be combined, eg, a sales ledger file containing reference data, such as name, address, account number, together with current transactions and balance outstanding for each customer.

Input files

These carry the input data to the system, via the input and validation program of the computer.

Transfer files

Transfer files (including change and sorted files) carry data from one processing stage to the next, particularly to update the master file.

Work files

Work files are intermediate transient files. They contain data selected from one or more files on which some form of process is required before the contents of the work file can be defined as a proper file; eg, the magnetic tape files required for a sorting process other than the input file and resultant sorted file.

Output files

Output files carry the output from a computer system, either to be printed, or for input to another system; eg, sales invoices to be sent to customers, or sales invoice transactions to be input to a sales ledger or accounting system. Information which is to appear as output must derive from data on one or more files.

Dump files

These are used for keeping a copy of the state of part of the system at a point in time. This may be a file which has been updated, a set of transactions which have been processed, or a program which has gone wrong. Dump files are basically used for security purposes.

Library files

Library files are used for storing application programs or modules or utility programs or software required by the computer system.

FILE STRUCTURE

Files can be considered to have a multi-level structure (fig. 10.1).

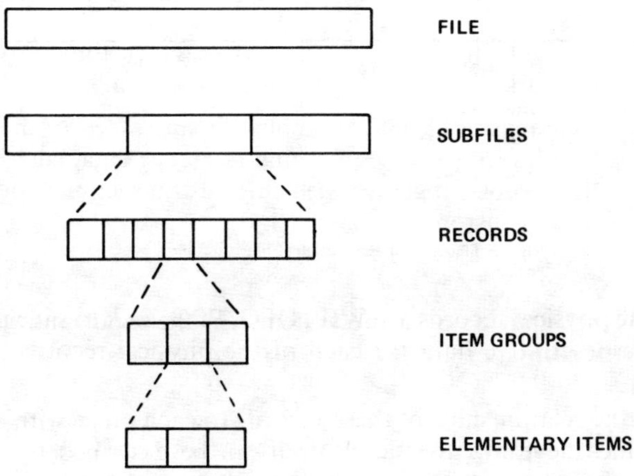

Figure 10.1 Multi-level file structure

Files and subfiles

A file may consist of various subfiles. In a production control system (fig. 10.2), the parts control master file may contain a series of records for each part manufactured, and each record could be considered to belong to a subfile in its own right, consisting of:

- work-in-progress records (WIP) detailing operations outstanding;
- outstanding parts order records (OPO) detailing orders outstanding for the specific part;
- purchase parts orders records (PPO) where the part is bought out.

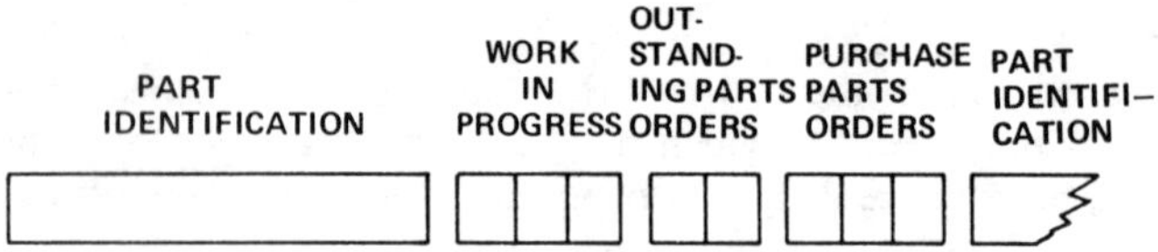

Figure 10.2 Parts control file records

The file structure (fig. 10.3), consists of a number of logically grouped records which have been physically separated for convenience of handling.

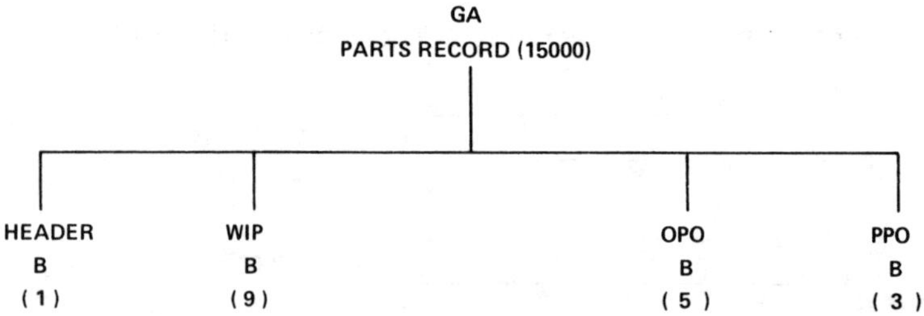

Figure 10.3 Parts control file structure

The separate physical records are WIP, OPO, PPO, and, to save repetition of the same descriptive data for each of the physical records, a header record.

To show the relationships of these records to each other within the file, the hierarchical levels of the family tree can be described by assigning alphabetic letters to each level, starting the letter A at the highest level, and adding the prefix G to indicate a group or logical record (see GA in fig. 10.3).

Recording file structure

As the elements of a file are defined they must be clearly recorded. The computer file specification (fig. 10.4) provides a suitable form. We see that at the highest level (fig. 10.3), this file comprises several groups of different types of record or subfiles, is given the name 'parts control' as a reference and level GA. This file contains 15,000 parts, each part therefore 'occurring' 15,000 times on the file. The entries in the occurrence column of the file specification form represent the number of times that a particular type of record occurs within the next higher level shown. In this case each record occurs at the same level (level B) within level GA. Each of the 15,000 parts comprises:

Computer File Specification

N C C

File description	System	Document	Name	Sheet
PARTS CONTROL	PC	4.4	PRODCON	1

File type: Input [ ] Output [ ] Master [ ] Transfer [ ]

File organisation: **Sequential (ascending)**

Storage medium		Retention period	Number of generations	Number of copies
Mag. tape [✓] Disc [ ] [ ]	Single [✓] Multiple [ ]	10 days	3	2

Recovery procedure

Keys: **Part number**

Labels: **Standard**

Level	Record name/ref.	Size	Unit	Format	Occurrence
GA	Parts control				15,000
B	Header	50	B	F	1
B	WIP	20	B	F	9 (av. 3)
B	OPO	15	B	F	5 (av. 2)
B	PPO	50	B	F	3 (av. 1)

Block/batch size: Actual, for fixed length	Block/batch size: Maximum, for variable length	Unit of storage	Number of blocks: Average	Number of blocks: Maximum
2000	–	Records [ ] Words [ ] Bytes [✓] Characters [ ] Cards [ ]	1550	3450

File size: Average	File size: Maximum	Growth rate
2,850,000	6,825,000	% per or determining factor

MAG. TAPE ONLY	Tracks	Recording density	Speed	Length
	7 [ ] 9 [ ]			

DIRECT ACCESS ONLY

Addressing/accessing method	Packing density	Frequency/condition of re-organisation
	%	

Level	Type of overflow	Size of overflow areas

Notes

S42

Author

Issue

Date

Figure 10.4 File specification

- one 50-byte header record;
- a maximum of nine 20-byte WIP records but which, on average, occur only three times;
- a maximum of five 15-byte OPO records, which occur, on average, only twice;
- a maximum of three 50-byte PPO records which occur, on average, only once.

The storage medium and the unit of measurement to be used need to be stated. In the example, the file is a master file to be stored on magnetic tape, so a tick is placed in the appropriate boxes. The unit of measurement is the byte and the letter B (in this convention) is entered in the unit column. The entry in the size column for each record is the actual size for a fixed record, or a range, minimum, maximum and average size for a variable record.

Records

Each of the records in turn consists of groups of data called data items, the arrangement and relationships of which combine to form the structure of a

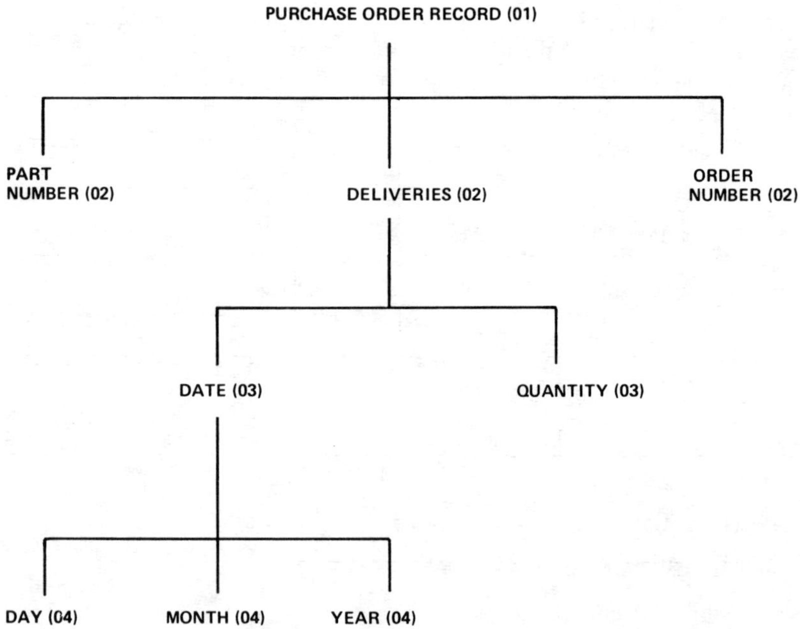

Figure 10.5 Item levels in a record

particular record. Figure 10.5 shows the data items contained in the purchase order record of the example.

It might be necessary to subdivide some of these items to still further lower levels. The lowest level to which an item can be subdivided is called an elementary item, in which case the upper level is called an item group. One way of denoting these levels is the one used in COBOL and PL/1. To each item is allocated a level number; an item at any given level includes all items following it until a level number less than or equal to it is shown. Many levels are possible, but in practice it is unusual to have more than five levels. Using this convention, level 01, which signifies the complete record, is not used. This subdivision is assisted by drawing a family tree diagram (fig. 10.5), examining each level and deciding if a further subdivision is necessary.

Specifying record structure

A description of the data items contained in each record must be clearly and methodically recorded; eg by means of the record specification form (fig. 10.6). The sequence in which the items are recorded represents the actual format of the record on its medium, the 'position' columns reflecting the position of each item within the record, in this example, in bytes.

The composition of each elementary item is described by the COBOL 'picture' convention. A picture is a symbolic representation of the format of a data item, and is fully described in the COBOL manual at an installation. The construction is:

- the position of the symbol in the picture indicates the position of the corresponding character or value of an item;
- the number of repetitions of the symbol indicates the number of repetitions of the character or value in the item. A number enclosed in brackets after a symbol may be used instead of repeating that symbol, eg A(4) and AAAA represent the same format.

An example of the picture symbols is shown below, but it is important that the COBOL version for a particular installation be used.

A A–Z or space

B inserted space

P assumed decimal scaling position

S leftmost character indicating an operational sign

V assumed decimal point

X any character

Record Specification NCC	Record description	System	Document	Name	Sheet
	Purchase parts order	PC	4.7	PURPRT	

Medium: Mag Tape | Record length: Fixed ☑ Variable ☐ | Record format: Fixed ☑ Variable ☐ | Record size: 50 | Words ☐ Characters ☐ Bytes ☑ | File specification refs.: 4.4/PRODCON | Lay-out chart ref.:

Ref.	Position From	Position To	Level	Name In system design	Name In program	Data Type	Size	Alignment	Picture	Occurrence	Value Range
1	01	05	02	PART NUMBER		C	5		Z(5)		1–99999
2	06	45	02	DELIVERIES						5	
3	06	11	03	DELIVERY DATE							
4	06	07	04	DAY		C	2		9(2)		01–31
5	08	09	04	MONTH		C	2		9(2)		01–12
6	10	11	04	YEAR		C	2		9(2)		01–99
7	12	13	03	QUANTITY		P	2		9(3)		001–999
8	46	50	02	ORDER NUMBER		C	5		X(5)		A0001–A9999

S 44
Author | Issue | Date

Figure 10.6 Record specification

Z 0–9: leading zeros replaced by spaces

9 0–9 (zeros not suppressed)

0 inserted zero

, inserted comma

. inserted full stop

* 0–9: leading zeros replaced by asterisks

£ £ sign: if repeated, indicates that it will appear to the left of the first non-zero digit, any zeros being replaced by spaces

+ If positive, plus sign, if negative, minus sign
– If positive, a space is inserted; if negative, a minus sign } If the symbol is repeated, the sign or space will appear to the left (or right) of the first (or last) non-zero digit, with all non-significant digits replaced by spaces.

E appears before the section of the picture relating to the exponent of a floating point value

Though the picture describes the format, it does not always correctly indicate the size of an item in terms of the space it occupies on its storage medium, which is influenced by the way in which the data is represented. In the example (fig. 10.6), an entry in the data type column of C indicates that the item is represented in character form and P in packed decimal (byte). Other types could include:

X hexadecimal 4 – bit code;

Z zoned or unpacked decimal (byte);

B binary (bit operations);

F fixed point binary;

E floating point binary;

The combination of the picture and data type enables the correct size (or space on the medium) of the item to be recorded according to its data type, eg, while 'quantity' consists of three characters, in packed form it is contained in two bytes.

Each item can be described further by indicating its value as a range (minimum-maximum) or in specific characters; this is used mainly for validity checking. An entry in the occurrence column indicates that particular item occurs more than once, and the identical items are not repeated accordingly.

Key field

One or more of the data items or group items will act as the key field of the record which is used for identifying the record for location and processing purposes. The key field for a particular file is shown on the file specification (fig. 10.4); in that example, the key is part number. In a sales ledger file the key field might be customer code; in a payroll file the key field might be employee number.

The term *field*, which is sometimes used as a synonym for data item, more specifically defines the space which an item occupies in terms of its data type. This is most easily conceived as the number of columns on a punched card occupied by a particular data item.

Record length

It is possible to store records in fixed or variable format and length.

Fixed-length records

The format and length of fixed-length records are always the same, each item being located in the same relative position within each record, and a fixed size being specified which is sufficient to hold the longest item of its kind. These are usually easier to design and program, but are more wasteful than variable-length records in storage space.

Variable-length records

There are several types of variable-length record situations (fig. 10.7):

- a group of different types of record, each of fixed but different lengths;
- records with a fixed minimum length and a variable number of fixed-length items following the fixed portion;
- records with a fixed minimum length and a fixed number of variable-length items following the fixed portion;
- records with a fixed minimum length and a variable number of variable-length items following the fixed portion;
- complete and random variability of length.

Often, it is possible to break down a variable-length record into a group of fixed-length records, so that with careful design, the advantages of both fixed and variable working can be achieved.

The use of several fixed-length physical records forming one variable length logical record, can make programming complex. However, it

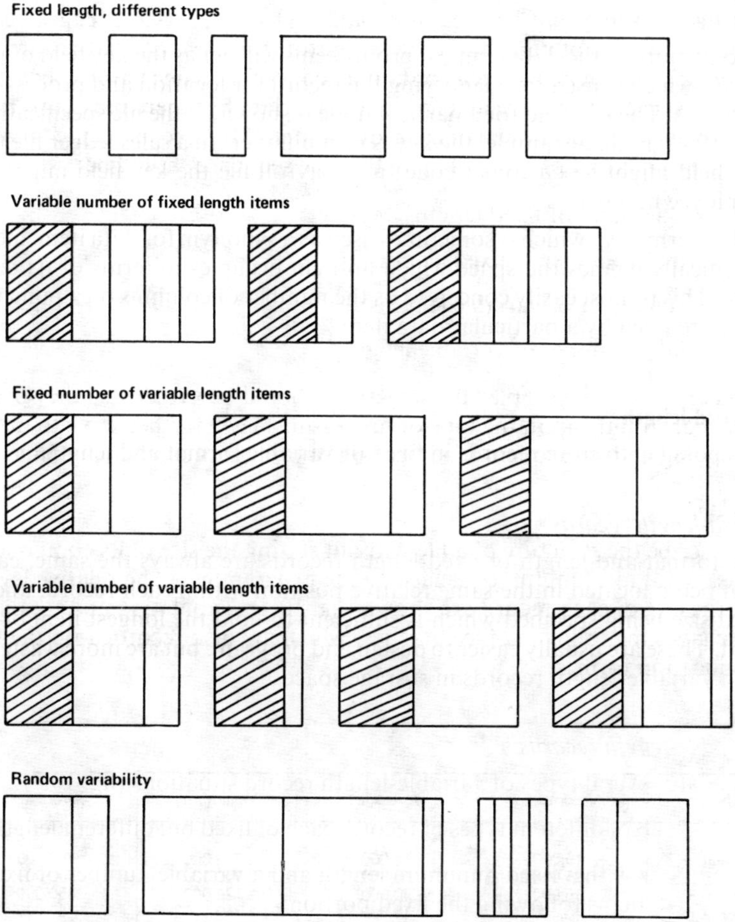

Figure 10.7 Variable-length record situations

eliminates the necessity to specify the maximum length of a record, and individual records can be processed one at a time, thus making the best use of storage. The production control example falls into this category; the parts control record which was defined as a group record can be considered as the variable-length logical record referring to the part. It will need to be variable because the various subfiles of work-in-progress, outstanding orders, etc, can vary in length. However, each individual record in the subfiles can be fixed in length.

Data items may also be fixed or variable in length; variable-length items must have an end of item marker or separator. If there is a variable number of variable-length items, each containing data of the same nature, then not only must each item have an end of item marker, but there must also be a different marker for the end of the group of items. It is recommended that if there are a mixture of types of item, that the sequence should be:

1 fixed number of fixed lengths;

2 variable number of fixed lengths;

3 fixed number of variable lengths;

4 variable number of variable lengths.

Where possible, complex file structures should be avoided as they often make constraints upon the type of processing. They further serve to make debugging and subsequent modification of programs exceedingly difficult.

File size

To describe the structure of a file without stating the size of its component parts, is not meaningful, especially to the programmer. The relative levels and the size of each record are shown (fig. 10.4). Tabulating and totalling these gives the following file statistics, which are also recorded on the file specification form.

	RECORDS		BYTES	
Name	Maximum	Average	Maximum	Average
Header	15,000	15,000	750,000	750,000
WIP	135,000	45,000	2,700,000	900,000
OPO	75,000	30,000	1,125,000	450,000
PPO	45,000	15,000	2,250,000	750,000
Total	270,000	105,000	6,825,000	2,850,000

Calculations such as these which depend on the number of variable records and the number of times one record occurs within the next higher level are normally estimates. When the file or its records already exist in some similar form then estimates of this kind can be soundly based on comparative established data, but if the file is part of an entirely new design, then the estimates are likely to be based on assumptions about related data and are in consequence less reliable.

Block size

Data can be transferred one record at a time between the backing storage device and main memory, but it is normal and more efficient to transfer data in blocks of records. The block size to be used with a particular file must be defined, and sufficient main storage assured to accommodate it.

The block size is influenced by the size of each record and whether they are fixed or variable in length. When a file contains only fixed-length records, then the block size is normally a multiple of that length; but when a file contains variable-length records, the block size is not less than the expected maximum record length and is normally a multiple of the average length.

There are likely to be other considerations to be taken into account when deciding the block size, eg software, other constraints on memory, and possibly a standing instruction from management that the block size used throughout the installation must not exceed certain minimum and maximum limits (say 1000 and 2000 bytes).

FILE ORGANISATION

The physical organisation of data on backing storage devices (eg magnetic tape/disks) can be defined as the relationship between the values of the key fields on consecutive records and the position on the storage device which they occupy. There are several different ways in which files can be organized, and for each a method of accessing the records must be defined. The choice of organisation method is often a compromise between the requirement for efficient maintenance of files (keeping them up-to-date) and fast retrieval.

The two requirements of efficient maintenance and fast retrieval are sometimes mutually exclusive, though not always; in any particular case, a trade-off must be made between one or the other. It is also generally necessary to achieve a practical balance between storage and processing costs of data. Processing costs can be reduced if the data organisation suits the characteristics and applications of the logical data. Storage costs may sometimes be reduced only at the expense of processing costs.

File organisation on serial storage devices

It is worthwhile to distinguish between the terms serial and sequential which are sometimes used as though they were synonomous and interchangeable. A serial file is one in which each data record is placed in turn in the next available storage space. The record keys need not be in any particular order, and there need be no relationship between the logical position of an item on the file and its physical position on the device. On a serial file, records can only be accessed in the order in which they occur in the given storage medium. This is contrasted with sequential file organisation where records are held and accessed in some predetermined sequence of the keys. The two expressions become synonymous when used for magnetic tape, which, because it is a serial recording medium, can hold records both serially and sequentially. Paper tape, punched cards and magnetic tape are serial recording media, differing from magnetic disks, drums and cards, which are direct access media. (Considered as a manual file, of course, punched cards may be a direct access medium.)

Basically, files may be organised on serial storage devices in one of two ways:

- randomly, where the records are not held in any particular order;
- sequentially, with the records held in a predetermined sequence.

For example, a transaction file of unsorted records held on magnetic tape may be organised randomly; after sorting the records to a particular key sequence, it would be a sequentially-organised transaction file.

A randomly-organised serial file has the simplest structure of all methods of file organisation. New records are simply added to the file at the end. In sequentially-organised serial files (sometimes known as serial sequential files) the records normally follow a certain sequence according to the ascending (or descending) value of a particular key field in the data record. Serial read operations will have the effect of reading the data in a structured manner, (ie sequentially according to some rule).

It should be noted that sequentially-organised files may be accessed in ways other than serially, and that files may be organised serially on non-serial media (eg direct access devices).

File organisation on direct-access storage devices

Direct-access devices provide an economically feasible compromise between unlimited main store and serial storage devices, by providing facilities for:

- storing large amounts of data on-line to the computer;
- accessing quickly any given piece of data on the file;
- transferring this data at high speed.

Direct-access devices comprise: magnetic drums, exchangeable and fixed disks, magnetic cards/strips, and 'floppy' disks. These devices have facilities which are not available on serial storage devices; for example, on disks:

- the read/write heads can traverse the whole extent of the device in a very short time;

- each block of data can be directly addressed and accessed by the hardware;
- the constant rotational speed is sufficiently accurate to allow the overwriting of individual blocks.

These facilities offer three different methods of organising files: serial, sequential, and random.

Serial organisation

A serial file is created by placing each record in turn in the next available storage space, leaving no gaps, apart from non-data areas between records. Records are entered in the sequence in which they are presented and there is not, necessarily, any regard paid to keys. Pages (or buckets) are fully packed, except that records are not split across page boundaries and therefore, packing density approaches 100 %.

Serial files on direct-access devices (eg magnetic disks) are treated in a manner similar to serial files on serial devices (eg magnetic tapes). Since there is no structuring, no room is left for insertion; and updating is by copying.

This type of organisation is not suitable for master files on direct-access devices where there is the need to insert key records into an existing sequence, or where access by address generation techniques is employed as there is no overflow facility. It is suitable for:

- transaction files;
- print files;
- dump files;
- all files where data is temporary.

The advantage of serial organisation is the maximum utilisation of space on the device. The time taken to find any given record is, on average, half the total time to search the whole file.

Sequential organisation

By definition, sequential files must be structured and stored in a certain sequence of a particular key item or group of items; it must be possible to access records directly by obtaining their location from the keys which identify them. (The sequence need not necessarily be physical. For example, if certain records in a logical sequence are displaced into overflow areas, they are no longer physically contiguous with the preceding and following logical records; they must be retrieved in sequence.) This method combines the advantages of a sequential file with the possibility of random access.

Sequential organisation is suitable for most master files in a normal batch processing environment. It is not normally used for fast-response on-line enquiry systems.

Different methods can be used for structuring and accessing sequentially organised files. These will be discussed in the section on file access methods.

Random organisation

A randomly-organised file contains records stored without regard to the sequence of their key fields. It is created by loading records in any convenient sequence and they are then retrieved by establishing a direct relationship between the key of the record and its address on the file. This can be achieved either by use of 'index' or by 'key transformation' (file access methods) to give the address of the record on the file.

Random files show a distinct advantage where:

- hit rate is low (ie number of records accessed in any processing run is small);
- data cannot be batched or sorted;
- fast response is required.

It is therefore the usual method of file organisation for real-time, fast response systems. Further advantages are that insertion and deletion of records is straightforward. The requirement to store in key sequence, imposed in sequential organisation, is removed. Deletion of records leaves gaps in the file which can be used for later insertion of further records, or alternatively insertions can be made at the end of the file.

It must be emphasised that when the terms sequential and random are applied to data, they refer to the way data is organised on a particular file and not to a method of processing. These terms, when applied to processing, usually refer to the sequence of the input transactions or the sequence of reference to records in the master file. It is possible to process sequentially both sequential and random files, or to process randomly both sequential and random files.

Catering for expansion (overflow)

A normal tendency for master files is to expand. Records may increase in size or records may be added. Even if the total size or number of records does not increase, there will almost inevitably be changes which will cause overflow.

It is usual to update files on direct access devices by overwriting existing data. At the same time there must be some provision for additions, and preferably some means of re-utilising storage arising from deletions.

Overflow can arise from:

- a record being assigned to an address which is occupied by another record;
- a record being expanded so that it can no longer be accommodated in its present ('home') area.

Overflow may be caused in a sequential file by attempting to insert a record in its proper place in an existing key sequence.

Overflow in a randomly-organised file can be caused by attempting to store more records than a given storage area can accommodate, or, in the case of key transformation, the existence of a *synonym* (ie where more than one key is transformed to the same address). There are three basic methods of catering for expansion:

- specifying less than 100% page packing density;
- specifying less than 100% disk cylinder packing density;
- specifying overflow pages.

The last two of these are sometimes called 'first'-level and 'second'-level overflow, respectively.

By specifying page-packing density, record expansion or addition of records will be accomplished automatically, using up available space. If expansion continues, however, there will eventually be no more space and alternative action must then be taken. 'Page overflow' has the major disadvantage that it will only be effective and efficient if expansion is regular. In cases where localised expansion occurs to any extent, (ie when the activity distribution is uneven), this system may be inadequate.

To cover the situation where expansion is less regular but still throughout the file, one or more pages can be left in each cylinder to cater for further expansion. During processing, any additions are placed in the overflow pages for the cylinder. If a large block of a sequential file suffers expansion, however, the combination of these methods may not cope, even for one run.

To cover the possibility of very irregular expansion in the short term, an area may be reserved for second-level or independent overflow. This area is reserved as a number of pages which are used as required. They would be used when all other overflow facilities have been exhausted.

File maintenance and reorganisation

It is not always possible to predict precisely how the contents of a file will alter; ie it is sometimes impossible to allow for efficient expansion. The use of first- and second-level overflow will always result in some loss of efficiency due to head movement and multiple accesses. The overflows will generally occur in high-activity areas and more time will be required for each access. Other areas will be empty due to low activity or deletion.

Even if expansion is regular and predetermined, to allow sufficient space for increase over a long period might be wasteful from the point of present disk usage. It is desirable, therefore, to reorganise at regular intervals, or, if the situation is unstable, as necessary. In the unstable situation, overflow provision acts as a temporary expedient and processing involves a cycle of a few runs followed by reorganisation.

In practice, a compromise is reached between the extra storage space and timing required to cater for expansion within the file and the time required to keep the file tidy by reorganising. Software which is used for file reorganisation can usually provide reports on the state of the file at any time, so that decisions can be made about reorganisation requirements.

Reorganisation will generally take place using magnetic tape as the intermediate medium, but another disk unit would be equally suitable. During the first run, records would be written in a reorganised sequence. They would then be written back from the intermediate medium to the file area. This, now becomes merely a matter of reloading without using overflow areas. Since index changes are usually required at this point, software will contain routines to reorganise indexes.

For a random file, where addresses are generated from the key, it may be necessary to amend the address generation method to take account of the impact of the expansion of the file.

FILE ACCESS METHODS

With serially organised files the method of access is quite straightforward because each record has to be read in turn from the device until the required one is found. The same applies to serial sequential files, though various techniques can be used to speed up the process of finding the record. However, with sequential and random files on direct access devices, other access methods are used, the main ones being indexes and key transformation.

Indexing techniques

Three types of index can be used: basic, implicit or limit.

Basic index

For each record on the disk, an index entry is held giving the key of the record and its location on the disk unit. To facilitate searching, the index itself is held in key sequence, although the actual records need not be. To locate a record, the index must be searched to find the required key and, thereby, the location of the record.

In order to make an insertion, it will be necessary to create a new index entry for the insertion and to insert this index entry into the correct place in the sequential index. This process can take a considerable time, it may be better to use a chaining technique within the index.

The example (fig. 10.8) shows the insertion of a record with key 10. The 'chain' pointer on index entry 4 indicates that the next index entry is number 101, and the chain pointer on entry 101 indicates that the next sequential index entry is number 5. (Index entry locations, rather than numbers would normally be used here.) After several runs of the job when several insertions have built up, it is advisable to arrange the index back into sequential sequence, to improve the efficiency of retrieval.

Index				File	
Entry Number	Key	Address	Chain	Address	Key
1	1	2	0	1	9
2	2	4	0	2	1
3	4	5	0	3	250
4	9	1	101	4	2
5	11	100	0	5	4
.	.	.	.	.	.
.	.	.	.	.	.
.	.	.	.	.	.
100	250	3	0	100	11
101	10	101	5	101	10

Figure 10.8 Basic index chained

When a deletion occurs on the file, the appropriate index entry can be cleared, either by using chaining to by-pass it, or set with a special record number to indicate that this key no longer exists on the file. In either case, the entry itself will be removed when the index is next reorganised. The deletions on the file itself will release space which may be used for later insertions and a separate list is maintained of available areas for use in this way.

This basic index method has the disadvantage of requiring large indexes and correspondingly high search times before a record is located, particularly if the index itself is held on the disk. For this reason, it is not very often used. If, however, a high activity portion of the file could be incorporated into the index, eg, items of stock balances which change most frequently, the majority of accesses to the main file could be saved.

Implicit index

With this method, an index entry is held for every possible key, giving the location of the record which bears that key or, alternatively, an indication that the key is not represented. The index is held in key sequence but there is no need to hold the keys themselves in the index. Since every possible key is represented, any entry in the index can be found implicitly from the key. An example is shown (fig. 10.9).

Index		File	
*Key	Address	*Address	Key
1	3	1	2
2	1	2	5
3	4	3	1
4		4	3
5	2	5	6
6	5	.	.
.	.	.	.
.	.	.	.
.	.	.	.
.	.	.	.
.	.	.	.

* Note that index key and file address would not actually be recorded

Figure 10.9 Implicit index

This method involves large indexes, particularly if there are gaps in the key numbering system, but index search time is small, since it is possible to go straight to the required place in the index. No restrictions on data record sequence are involved, since every record is indexed and no difficulties are caused by insertions or deletions.

This method is the fastest, although not always the most compact method of indexing a random file. Its compactness can be increased at the expense of access time by closing-up known gaps in the index.

Limit index

Unlike other indexing methods, this method implies some degree of ordering of the data itself, and so cannot be used for indexed random files.

With this method, the data is divided into pages in such a way that each page contains a range of keys. If the file was in sequential order, the range in one page would not overlap the range of keys held in any other page. The highest key in each page is then indexed.

The limit index provides a useful way of reducing index size and is the normal method of indexing used with an indexed sequential file.

Indexes may exist at a variety of levels, but for files contained in a single disk pack they would normally comprise:

- cylinder index;
- track or page index.

Usually, a cylinder index would be written on the first cylinder of the file. The track or page index would be written on the first track of each cylinder.

When a job begins, the cylinder index will be read into main store if software allows this, for the duration of the run. The track or page index for a particular cylinder is moved into core either wholly or in parts as required. The advantages of this method are that it is possible to select appropriate blocks of records without reading the whole file and that overflow is generally supported, thus giving a higher packing density.

Key transformation techniques

Address generation methods allow information to be held on the disk without an index and without being tied to serial processing. The principle is that some operation (algorithm) is performed on the key which will generate a unique address for each different key.

The first and most obvious method is to use the key number itself as the address. The disadvantage of this is that in the vast majority of cases gaps in the addressing will result, owing to non-existence of certain keys. These gaps can cause wasted space upon the disks, and must be closed.

Key transformation is a development where relative addresses are derived mathematically from the value of the key itself. The advantages are that: as indexes are not required, space and searching time are saved; insertion and deletion of records can take place indefinitely without file reorganisation and thus files can be on-line for 24 hours per day; there is no restriction on processing, which may be sequential or random; and a relatively small programming change is involved if the formula has to be changed.

However, there are also disadvantages. Variable-length records are difficult to handle; often, the generated addresses are not in key sequence, which can create difficulties for serial processing (it may be necessary for the input to be sorted to the address sequence); gaps in keys can cause wasted space; synonyms can occur where several keys give the same address; and allocation of efficient overflow areas to a randomly-organised file is difficult (it could lead to large areas of the disk being empty with overflow areas being set up for other areas of the file).

In designing the algorithm, the main aims are to eliminate large gaps in the distribution of keys and to minimise the occurrence of synonyms. The extent to which this will happen will be governed by:

- the distribution of keys;
- file density;
- effectiveness of the algorithm;
- the number of records per block or per page.

The principle of all formulae used in address generation is to provide an even spread of keys through the range of generated addresses. This avoids, as far as possible, the situation where some pages are comparatively empty, and where others are overflowing due to the occurrence of synonyms. There is an element of trial and error in selecting the best algorithm but the following steps can be taken to assist in selection:

- analyse the pattern of present and future codes;
- build in constants to over-ride gaps in the structure;
- call on the experience of manufacturers and other users;
- test suggested algorithms, using simulation;
- determine which gives the best results.

Standard methods which are used are described:

Division-taking quotient

Assuming that the keys are in the range M to N and that the file is divided into pages (P), the steps are as follows:

1 subtract M from the key, giving a number in the range of 0 to $(N-M)$;

2 calculate the average number of keys per page, by dividing the total number of possible keys by the number of pages $\left(\frac{N-M+1}{P}\right)$;

3 divide the number derived from the first step by the number derived from the second step;

4 ignoring the remainder, take the integral quotient to be the page number.

This method is suitable if the keys are already evenly distributed through the range of possible keys. As the file is generated, the pages are filled in sequence.

Division-taking remainder

The steps are:

1 divide the key by the number of pages;

2 ignore the integral quotient, take the remainder to be the page number.

This method gives better results if the keys are unevenly distributed. It does not give good results if gaps appear evenly in the key numbering and if the interval between these gaps shares a factor with the number of pages.

Truncation

This is really a special case of division that is used if the divisor can be made a multiple of ten or of the radix for non-decimal keys. Truncation of digits from the right of the key is equivalent to 'division-taking quotient'. Truncation from the left is equivalent to 'division-taking remainder'.

The two may be combined by truncation from both left and right and accepting the centre digits. In this case, the process is called *extraction*. Truncation requires caution and often does not give satisfactory results if used on multi-part keys, where different sets of digits have particular significance and the full range of values for each set is not used.

An example of the use of truncation might be, with a product file whose keys range from 10001 to 89999, to truncate by one digit from the left. This would give a new key range of 0001 to 9999. Some numbers would give the same result (for example, numbers 21000 and 81000 would each give 1000); in fact, a maximum of eight synonyms is possible for each key. The records may now be stored in relative pages 0001 to 9999, the number of records per page being determined by the density of the data; if only 20,000 records exist (ie 25% of the total possible records) the blocking factor may be 25% of maximum synonym size (ie 2 records per page). Overflow will clearly be a problem with this method.

Folding

In this method the key is split into two or more parts which are then added together. Truncation may then be required to bring the result into the required range of page numbers. For example, the key 891473995 may be folded by splitting into three groups of three digits, and adding the groups:

$$\begin{array}{r} 891 \\ 473 \\ 995 \\ \hline 2359 \\ \hline \end{array}$$

This would give the location of the record as page 2359.

Radix conversion

In this method the key is treated as a number in some radix other than its own and is converted to its own radix. For example, a decimal number might be treated as in radix 11 and converted to radix 10. Taking the key 59500 as a radix 11 number, the radix 10 equivalent is 943679, a number which may be suitable for further reduction, using other techniques, to give a suitable page number.

Squaring

Many methods are based on squaring, usually in combination with one or more of the previously mentioned methods, for example:

- squaring and extract;
- squaring, fold and extract.

Of all the above methods, the one which generally gives the best results is division by a prime number, taking the remainder. Folding also often gives satisfactory results. Radix conversion and squaring are not really recommended unless there is some particular advantage, due to the nature of the keys. Any address generation formula chosen should be validated empirically against the actual set of keys to be used before a decision is made.

It is possible to organise files on a random principle yet retain the ability to process sequentially after a single random seek. This will occur where groups of records need to be sequentially processed within any one group. The method adopted is to split the addressing method into two parts:

- use an algorithm on the senior digits of the key to provide the address of the starting point;
- add the junior digits to obtain sequential record addresses from that starting point.

Addressing for overflow

There are two basic methods of addressing overflow areas in sequentially organised files: chaining and tagging.

Chaining

In this method, the principle is that whenever overflow occurs, a reference is written as a guide to the page or track in which overflow is contained. In the case of a sequential file, this would take the form of a record within the page and would give the overflow page address. This address could be within the same cylinder or could be within an area dedicated to handling overflow for the whole file.

For an indexed sequential file, as soon as overflow occurs an additional index entry is written in the appropriate part of the track index. Figure 10.10 shows the highest key to have overflowed from this prime track and the address of the lowest keyed record to have overflowed.

CHAINING

Before overflow

Track 0001

001	002	003	006	008	012	

Track Index		Overflow Index	
Address	Highest Key in Track	Address of Lowest Key	Highest Key Over-flowed
0001	012	0001	012

Overflow index identical to track index before overflow occurs

CHAINING

After overflow following insertion of records 009 and 011

Track 0001

001	002	003	006	008	009	011

012 Overflows to track 0031

Track Index		Overflow Index	
Address	Highest Key Over-flowed	Addr. of Low-est Key Track/ Record No.	Highest Key in Track
0001	011	0031	012

Figure 10.10 Chaining for overflow

Each record in the overflow area is chained to the next highest key by a link field, giving the track and record number: the overflow from a given prime track always has its own chain within the overflow area.

Tagging

Chaining involves detecting a record by following the chain of addresses. Tagging follows a different principle, by leaving a remainder in the form of a tag for every record which has been displaced. The tag, which is written by a housekeeping routine, contains the record key and the address to which it has been written.

Tagging has the advantage that every record can be directly traced without the need to follow a chain. It has, however the following disadvantages:

- a data record may need to be displaced to make room for the tag, thus requiring a tag for itself.
- second-level overflow can be reached when:
  - the 'home' page is filled with tags;
  - overflow pages or all pages in a cylinder have been filled.

A further area at the end of the file has to be reserved to deal with second-level overflow. This consists of extension pages specifically related to 'home' pages and chained by a pointer in the 'home' page. Thus the concept of chaining enters the tagging environment.

FILE DESIGN CONSIDERATIONS

The choice of the most suitable data organisation for a particular group of applications is important, perhaps even crucial, to the eventual successful performance of systems which use the data. Ideally, the range for choice of storage medium and data access method should be unlimited, but this is rarely the case. The file designer is constrained by various factors, including the requirement to interface with existing applications, their files' existing data storage media, and the range of options supported by the incumbent manufacturer's hardware and software. It is, however, possible to draw up a series of alternatives in designing files, and to evaluate those alternatives against given criteria. The first consideration is the choice of data storage device to be used.

Choice of medium

Choice of the correct device for storing a file is most important with master files and most of this section is concerned with them, but careful consideration should also be given to media for other files. These are some of the considerations.

Purpose of the file

Direct access devices, such as disks, can be used to advantage as a means of holding transaction data or intermediate files; serial reading and processing from disk may be quicker than reading from magnetic tape. Although transaction and intermediate files on disk can be sorted quickly and conveniently, it is also possible to load transactions to specific areas of the disk, using, for example, data type as the criterion. Specific areas may then be accessed as required, thus removing the necessity for a sort.

Disks as well as tapes can serve as a convenient method of holding output data. This is particularly true, if a number of differently sequenced reports are required from different parts of the file. Random or selective sequential processing might be used for their production.

Program library files may be held on tape or disk. Frequently used programs are best held on disk, as they can be loaded without wasting time in searching through a tape.

Availability of hardware

Where there is no possibility of additional hardware for a proposed system, the choice of file media is restricted, particularly in small installations. It may not be possible to put the file on the best possible medium; the systems analyst will have less difficulty deciding which medium to use, but more difficulty in designing procedures which meet the needs of the user.

On the other hand, if major expenditure on hardware is envisaged, the systems analyst's choice may be more difficult, and new techniques may have to be developed. Not only must the optimum requirements of the system be considered, but also the requirements of other systems which may need to use the hardware.

Method of access

One of the first questions to be asked here is 'is direct access an essential requirement of this system, or is any future system likely to require this file?' If the answer is 'Yes', then choice will be restricted to the various direct-access devices available. Consideration must be given here to present and future requirements for interrogation of the file and multi-access systems. If the proposed systems require real-time operation, or are on-line in other ways such as for time-sharing or enquiry, then direct access will be essential.

If the answer is 'No', the choice between serial and direct access will be made on a consideration of timing and cost. Even if direct access is not a prerequisite, the use of direct access media may be desirable if this will reduce sorting time, enable the file to be split into smaller files, or significantly reduce the processing time.

File activity

Other basic questions to be asked are:

- how often is the file required for reading or updating?
- will it be used by more than one program? If so, will the sequence of operation of these programs have any significance?
- what percentage of the file will be required during the run?
- how is the activity distributed through the file?

The frequently-used file is less likely to tie up hardware. Additionally, it will require less operator time if it is held on a direct-access device. If serial processing from magnetic tape seems to be indicated by the characteristics of the main run, the implications for subsidiary runs must also be considered.

File size

Small files are usually best kept on disk, because they can be allocated sufficient space without affecting the rest of the disk. The same file on magnetic tape might use only the first 30 feet of the tape, so that the remainder cannot be used or can only be used for several other small files.

If a process requires many small files, then on magnetic tape:

- a large number of tape decks may be tied up;
- time can be wasted in searching for subfiles;
- time can be wasted in changing tape reels;
- files may need to be processed in sequence.

If the files could be stored on direct-access devices, they could possibly all be kept on one or two cylinders of a disk without incurring any hardware or time penalties.

Another possibility in the case of small files, is the use of slow media, such as cards or paper tape. This can be particularly useful when the file is only required at infrequent intervals, as cards and paper tape are cheaper as storage media, than disk or magnetic tape.

Large files are usually held on magnetic tape as it is a cheaper storage medium than disk. A large magnetic tape reel can hold more data than a 10-surface disk pack. Tape reels can be changed more quickly than disk packs when dealing with a multi-volume file. Furthermore, it is easier to allow a file to expand on magnetic tape, than on disk, particularly when the expansion involves an extra reel or cartridge. Other criteria may outweigh these considerations, however. It may be worthwhile to hold a large file with a low hit rate on disks.

A medium-sized file is considered to be one with the range of 4 to 40 million characters. This presents a rather more difficult choice: the particular hardware available may be the overriding factor.

Output requirements

If any one output report requires information from every record on the file, then this should be produced by serial processing. But if a report requires information only from certain parts of the file, it may well be better to hold the file on a direct-access device.

Where multiple reports are required from the same file, they may be produced either by direct access for one report at a time or simultaneously by serial processing techniques. Often some further processing may be required between extracting the information from the file and printing it; for example, sorting or conversion of data from one format to another, the need to transmit it over a data link.

Input requirements

Here it is important to choose carefully, not only the medium for the master file, but also the medium for input. For an input transaction file, requirements to be considered are:

- validation;
- control;
- sorting.

It may be necessary to separate different transaction types, to retain valid transactions whilst invalid transactions are corrected, to sort transactions to the same sequence as the master file. Thus the choice of medium for both transaction and master file may be interrelated.

There is also the case where the same input data, either in part or in total is required for two or more processes. If possible, the process of reading and validation of data from a slow peripheral should only be carried out once. The need for sorting this data several times must then be balanced against the possibilities of direct access.

Furthermore, one must consider whether to have a single transaction file or a number of files holding different types of transaction. In the second case, disks would be more attractive for holding the large number of small files.

Sorting implications

Sorting is a time-consuming process, possibly the most time-consuming aspect of file processing. Its elimination or minimisation can make significant savings in the use of time. The use of multiple files for different types of transactions is worth considering. Another possibility is the splitting of records between files. It will usually be quicker to sort two files with short records than one file with long records.

The available software must be considered. It is difficult to generalise, but the efficiency of sort packages varies and it may well be that a specific manufacturer has better software for disk than for tape (or *vice-versa*).

Hardware also enters the picture, particularly where large files have to be sorted. Given eight or more magnetic tape decks, tape sorting is attractive, but with only three or four decks, it can be a slow process in comparison with the use of direct access.

Speed of processing

We must be sure that the device selected is used to its best advantage and that as much simultaneity as possible is achieved. Points to be watched are:

- data transfer rate;
- latency and seek time on disk;
- tape passing and stop/start time;
- data block sizes;
- available working storage and the use of double buffering;
- the efficiency of software.

Cost

The cost of storage is not always a function of the cost per reel of magnetic tape, or the cost of a disk pack. Tape is obviously cheaper for storing large files, but not necessarily so for small files. Forty small files, each on separate reels of tape might easily go on to one disk pack. Although a disk pack is around ten or twelve times dearer than a reel of magnetic tape, six hours of unproductive tape searching could well compensate for that difference.

It is important, also, to remember the cost of keeping copies of files and the physical cost of storage space. If interchange of data between installations is an important factor, transport charges may also need to be borne in mind. Tapes can be posted more easily than disks which are bulky and heavy.

Choice of file organisation method

When choosing the file medium, some consideration will have been given to the way in which the file is to be organised. If the medium chosen is magnetic tape, then the organisation by definition must be serial. If, however, a direct-access device is chosen, a variety of organisations are available. A number of factors will affect the choice of file organisation.

File activity

This may be defined as the number of different records accessed on a run, divided by the total number of records in a file. Its percentage, distribution and amount will have a significant effect on the choice of file organisation.

If a low percentage of the records are to be processed on a run, the file should probably be organised in such a way that any record can quickly be located, without having to look at all the records on a file. They will need to be accessed directly, using one of the methods described earlier.

With certain methods of organisation, some records can be located more quickly than others. The records processed most frequently should be the ones that can be located most quickly. Where records are addressed directly, it may be preferable to load them in sequence of activity, so that the most frequently accessed records get the first chance of going to the page determined by the algorithm. Indexed sequential processing is satisfactory for high activity files and if the activity code can be included as the first digit of the key, then high activity records will be stored adjacent to each other, thus helping to minimise head movement.

An active file, that is one which is frequently referred to, must be organised very carefully, since the time involved in locating records may be considerable. At the other extreme, an inactive file may be referred to so infrequently that the time required to locate records is immaterial. A balance must be struck between the total amount of access time required if direct or indexed sequential approaches are adopted, and the amount of sort time required for presenting transactions sequentially.

If transactions are to be sorted, the hit rate may be such that one record in each block may require processing. If this is so, then a non-selective method of processing, producing a new copy of the file, may be worthwhile. If the hit rate is not such that one record in each block requires processing but the hit rate is still high, then non-selective sequential processing may still be used. For lower hit rates where several blocks per cylinder are likely to be accessed, indexed sequential processing can be used.

To appreciate the effect of file activity, consider a large file which has 90% of its records updated daily. Serial processing of a serial file is reasonable and is common practice in these circumstances. The transactions are sorted and serially matched against the master file, ensuring the maximum hits per file pass. If the percentage of activity drops to, say 70% or 50%, then this system would operate less efficiently. In general, as the percentage of activity becomes smaller, serial processing becomes less efficient. With a 90% activity, we are at one end of the spectrum; with only two transactions a day, we are at the other end. As file activity increases in a serial system, total processing time reaches a maximum. In a direct access system, total time increases, except that for an efficiently organised sequential file significant compensation can be achieved by having a higher hit per access.

Under given conditions of file activity and average hits per access, there will exist a break-even point where the total processing time in a serial system is equivalent to that in a direct-access system. Just where this lies for any given application depends upon many factors, but will typically lie in the file activity range of 5% to 10%.

File volatility

This term refers to the rate at which records are added to or deleted from a file. A static file has a low percentage whilst a volatile file has a high percentage. No matter how the file is organised, additions and deletions are of significant concern and can be handled more efficiently with some organisations than with others. The directly addressed file presents no problems other than where overflow is concerned. The indexed sequential file will require chaining of indexes to cater for additions and deletions. This will generally be handled by software. A serial file will require processing by copying.

File size

A file so large that it cannot all be on-line at one time must be organised and processed in certain ways. A file may be so small that the method of organisation makes little difference, since the time required to process it is very short no matter how it is organised.

Generally, a multi-volume file that cannot all be on-line at the same time, could not be processed randomly; a sequential file would have to be processed sequentially; a random file could only be processed serially; if the file is partially indexed, further levels of index would need to be established.

The growth potential of the file must also be considered. Usually, files are planned on the basis of their anticipated growth over a period of time. Initial planning must also consider how a growth that exceeds this will eventually be handled.

Data format

The physical arrangement of data on both magnetic tape and disk has been discussed earlier. There are, however, a number of more detailed points to be considered when designing disk files.

The density of packing or page utilisation chosen, and the number of seek areas employed, will have a significant effect on timing. There may be hardware or software limitations on page and block size. The factors to be decided upon, when specifying packing density are:

- provision of sufficient room to allow insertions and expansion in the home page;
- density not so low as to waste space on the file.

When packing for a serial or sequential organisation, it is quite reasonable to aim for a high level of packing, near 100%. This is particularly true if the file is to be processed by copying or is only a temporary file.

If the file is to be organised on an indexed sequential basis, the question of overflow arises. Theoretically it would be possible to load such a file with a high packing density. This is an acceptable arrangement if the file has low volatility. If there are many additions, records will overflow into embedded or independent overflow areas.

The effect of this will be to increase processing time, and particularly if independent overflow is used a high packing density could create an unwieldy system.

When packing a file for direct addressing, it is necessary to allow space for all possible keys, so that 100% packing will be useful. Many of the allocated record spaces may, however, be redundant.

Using a randomly organised file, packing density will become increasingly critical if the distribution of keys is uneven, or if the algorithm fails to smooth out this uneveness: a high level of synonyms will be produced. The systems analyst cannot specify the packing density on loading as when creating a sequential file; however, the same effect can be achieved by allocating more room than is required. If individual records are addressed, then every synonym will produce overflow, but the number of synonyms will be less. If tracks or pages are addressed, the systems analyst can specify the packing density of those tracks or pages, and this in conjunction with page size, will give a guide to the amount of overflow to be expected.

Where overflow does not arise or is insignificant, the higher the packing density the faster the processing, particularly with sequential processing. If overflow is present, the higher the packing density the greater is the overflow and the slower is the subsequent processing.

There is thus, a necessity to strike a balance between time savings (resulting from high density packing) and time losses (caused by high overflow percentages).

Blocking

Dependent upon the manufacturer's approach, one can either have variable-length blocks within fixed-length tracks, or fixed-length blocks, multiples of which make up various-length pages. In the first case, a block may only incorporate a single record. Whereas on tape, blocks can be as long as core availability allows, on disk, block size has to be related to the amount of space wasted where a multiple of the block size does not fit conveniently into the fixed-length track.

The advantages of blocking are that the utilisation of storage can be improved and there is a faster transfer of data. Each record may need to carry a certain amount of control data. This will only be required for the block, if records are grouped into blocks. The time saved on direct-access devices is not with inter-block gaps, but the rotational delay. If records are blocked, rotational delay exists only between each block or page.

Problems exist with blocking on direct-access devices. If records are to be processed sequentially, and the hit rate is likely to be quite high, then there are obvious advantages in reading a number of records into store at once, and searching for the appropriate record in main store. If, on the other hand, the hit rate is low, blocking may be a disadvantage, since it takes longer to transfer and unpack a block rather than the single record which is to be processed. Indexed sequential processing might be employed under these circumstances, although, here again, blocked indexed sequential is not unusual.

Blocking on randomly organised files may defeat the object of this type of organisation. Using direct addressing, the physical position of the desired record can be calculated quite easily. It is still necessary to transfer the block, but the record position in main store is known without the need to search.

File density

The density of a file is the proportion of the space on the storage device allocated to the file which is actually used by records in the file. A sequential file normally has a density of 100 %, where :

$$\text{Density (\%)} = \frac{\text{number of data characters}}{\text{number of character positions available}} \times 100$$

In general, the greater the density of a file, the lower the storage costs.

File growth

If the size of the file is to increase steadily, then the data organisation must be designed to accommodate, efficiently, new data records. With sequential files, the file size is not a particularly constraining influence; whereas, with other data organisations, space for growth must be provided within the total area occupied by the data.

Frequency of maintenance

Frequency of file maintenance is the number of updates to the file in a certain time-period. There are two philosophies where maintenance is concerned; transactions may be withheld until an economic-sized batch is ready for processing, or transactions may be processed as and when they arrive, ie, real-time. For batch systems, high frequency of maintenance may mean that maintenance processing is a major cost. Real-time systems may require batch maintenance during off-time periods.

SUMMARY

The major task in physical design of the computer subsystem is the design of files, and this chapter has presented some of the considerations involved. It is important to realise that the design of the physical system is totally interrelated and so file design cannot really be considered in isolation from inputs, outputs, procedures, codes, security requirements etc, all of which will influence the design of files. The systems analyst's major effort on file design will be concerned with master files, but other types of file need to be carefully examined. The file has to be physically structured in the light of the logical structures discussed (chapter 9, vol. 1); it has to be physically organised on the appropriate device; and methods of accessing the data have to be determined. The various considerations to be taken into account in selection of the medium, the file organisation method and the file access method are interrelated, and file design decisions require considerable experience on the part of the systems analyst.

11 Database Design

INTRODUCTION

The development of database systems is not a major breakthrough in computer technology; rather it is a logical development in the methods used by computers to access and manipulate data stored in various parts of computer systems. Thus this chapter is a logical development from the previous one; the systems analyst who is involved in designing databases needs to be fully conversant with the concepts and techniques of file design described there.

The overall objective in the development of database technology has been to treat data as an organisational resource and as an integrated whole. Database management systems allow the data to be protected and organised separately from other resources (eg hardware, software and programs). Defining the term 'database' is difficult. If it is defined as 'an integrated collection of data', then database systems can be found with very little integration of data. The definition which includes 'without duplication of data' is matched by the data base system with many duplications. It can be argued that the 'database' has always been there, and it has only been given a new name; to some extent this is true. Genuine database systems are marked by an approach to data management which is an entirely separate function from application programming (ie, a distinction between data management and data manipulation). Computer-based systems which support a centralised data management function can support a database. If the data management software can provide centralised access to the data from the programs, it is possible to treat the data as a separate resource.

Usually the centralised data management software is a software package called a Data Base Management System (DBMS). Within it will be found components which exist as separate software packages, such as a complex file handler, an information retrieval system, or a comprehensive report generator. But a DBMS will also contain components which are not found in other data management packages, such as recovery facilities, privacy controls and a specific data accessing language to be used by the programs. The most significant difference between a DBMS and other types of data management is the separation of data as seen by the programs, and data as stored on the direct access storage devices. This is the difference between *logical* and *physical* data.

This chapter continues by examining some of the objectives of database design, the approaches to logical and physical data, methods of organising physical data, and finally the database design process.

DATABASE OBJECTIVES

The organisation of data in a database aims to achieve three major objectives: data integration, data integrity and data independence.

Data integration

In a database, information from several files is co-ordinated, accessed and operated upon as though it is in a single file. Logically, the information is centralised; physically, the data may be located on different devices and in widely scattered geographical locations, connected through data communications facilities. In order to achieve the objectives of data centralisation, links between data must be maintained; direct-access techniques are used to permit efficient and flexible linking, although sequential organisation can be used in a database.

It must be possible to access data records using a wide variety of search keys. This can reduce the costs of implementing new applications, where multiple references are made to the same data, eg order processing of home, overseas and government orders at various factories. Data integration is achieved by using the techniques described later, chaining, data inversion and indexing.

Data integrity

Very often, within the same computer system, reports or analyses referencing the same logical information are inconsistent owing to differences in duplicated physical data. This could, for example, occur when changes are made to data in one file but not to a copy of the same data in another file.

One way to solve this problem is to ensure that when a field is updated, all other copies of that field are updated at the same time. This becomes difficult when copies of the field are held in separate files which are used by separate programs. Another way to solve this problem is to store all data in one place only and allow each application to access it; this is the database approach. This approach to data integrity results in more consistent information; one update being sufficient to achieve a new record status for all applications which use it. This leads to less data redundancy: data items need not be duplicated. There is also a reduction in the direct-access storage requirement. It will probably not be possible to achieve complete non-redundancy of data, due to performance, security and back-up requirements, but this approach makes it possible to control and maintain redundancy at minimum levels.

Data independence

Applications evolve as more information is required, and its usage improves. Changing requirements will influence the need to use stored data differently. Conventionally, this results in a certain amount of modification to the physical data to meet these changes, with perhaps an increase in data duplication and redundancy.

Conversely, when control and optimisation of the system are improved by the reorganisation of physical stored data and the installation of new hardware and software, existing applications must recognise the new physical data organisation; and so modifications to application programs are necessary.

Data independence is the insulation of application programs from the changing aspects of physical data organisation. This objective seeks to allow changes in the content and organisation of physical data without re-programming of applications; and to allow modifications to application programs without reorganising the physical data. The DBMS will match new and different programs to the data by relating their subschemas to the existing schema; new physical data structures will be accommodated by cross-referencing the new organisation to the existing schema. The concepts of schema and subschema are described below.

LOGICAL AND PHYSICAL DATA

The concepts of logical and physical data can be illustrated by a set of filing cabinets, a filing clerk, and some departmental managers who use the files. Physical storage is represented by the filing cabinets, the units of physical storage being the cabinets themselves, their drawers, the suspension folders, and the pieces of paper they hold. Physical data is that which is written on those pieces of paper. Let us assume that the filing system supports a motor insurance activity, and the data is concerned with policy holders, claims and other such information. One manager who is concerned with handling new claims instructs the filing clerk to retrieve certain data, such as customer details and previous similar claims from the files.

The filing clerk, who understands the structure of the filing system, retrieves the required data, and presents it to the manager in accordance with the instructions; perhaps, in a certain sequence in summary form after having made calculations based on some of the retrieved data. This is the logical data for that department manager. Other managers may require the same data, but with different instructions and results. In this example, the filing cabinets are equivalent to direct-access storage devices, the managers are programs, and the filing clerk is the DBMS.

If a new filing system is introduced, the only retraining necessary is that for the filing clerk; the managers may view the data in the same way as before. When managers wish to change their method of working, or perhaps a new manager is appointed, the filing clerk is informed of the new logical data requirements and uses the same physical data as before to fulfil them.

A manager never retrieves data directly from the filing cabinets. Because the filing clerk is the only person who knows how the data is organised and how the filing system works; the filing clerk is the only means of access to the physical data.

In a computer-based data processing system, separation of physical and logical data provides the same advantages as in our filing cabinet example. Application programs may be restructured or replaced without damage to the data. The data can be re-organised on the same storage device or new device types without change to the programs. This is because the DBMS is able to interface logical and physical data, and make the necessary adjustments when one or the other change. Because the DBMS is the only way of accessing the data, privacy, recovery and integrity controls are easier to implement.

Database terminology

In order to promote the portability of software among computer systems, standard terms are often employed, particularly in standard compilers such as COBOL or FORTRAN. This in itself does not guarantee portability but at least allows a common basis for design and conversion. Beyond the realm of compilers, very little has been achieved in the field of standard terminology, and this is true of database systems; different systems are often unrelated and use entirely different terminologies.

However, the Data Base Task Group (DBTG) of the Conference on Data Systems Languages (CODASYL) has proposed specifications for non-manufacturer-oriented terminology for database systems. (Details of CODASYL recommendations are given in Appendix B.) The specifications have not been used in all DBMS implementations, even those which claim to follow the CODASYL recommendations. A wide variety of terms to describe the nature of logical and physical data are used but this section will begin by explaining the CODASYL terms.

Logical data is described in terms of DATA ITEMS, DATA AGGREGATES, RECORDS and SETS. Physical data is described in terms of AREAS and PAGES.

Logical database units

In the motor insurance application example, the customer details could be a RECORD consisting of a customer's name, address, vehicle details, details of previous claims, account number, policy numbers. When the details of a particular customer are required by the manager, the manager gives the filing clerk the customer's name, and the filing clerk takes the record from the files.

The customer name in this case is a KEY; if the files are organised in alphabetic sequence of names, then this is a PRIMARY KEY. If the filing clerk has first to match the name, in a card index, with a reference number, by which the files are organised, then the customer name is a SECONDARY KEY, and the reference number the primary key.

The contents of the record are generally grouped together either on one sheet of paper, or on several sheets of paper clipped together or in a folder; if a particular customer's record contains many details, it may be necessary to use a second folder. When a customer is a large organisation with numerous policies and vehicles, then only recent data may be stored in the main files, other less frequently used data being removed to separate filing cabinets, perhaps in a separate location such as an archival store. The record may thus be retrieved in part or in whole.

In this case, the manager should specify the parts of the record required; for example, a manager requiring details of customers' accounts may not also need details of vehicles or policies. The filing system can accommodate this, and make life much easier for the filing clerk, by abandoning the record format which contains all customer details, and storing separate records for each client's name and address (eg header information), policies, accounts, and claims. Under these conditions, if the manager who handles claims requests the details of a certain customer and past claims, this may be done by first including the claim numbers on each client's header record. The filing clerk then matches and retrieves the client's header record, notes the claim numbers, then matches these numbers against the separate claim records file and retrieves the detailed claim records.

Alternatively, the relevant claim records may follow the client header records in the files. In this case, the manager is referring to an association between different record types, this association is called a SET. The filing clerk is fetching an OCCURRENCE of the set. Another set could be a client header record, accounting records, and policy records; here, there are three record types (there is no particular limit). A set must have a unique identifier or primary key, this means that one of the record types in the set has a primary key which is also used to identify a particular occurrence of the set. This record type is the OWNER record type of the set and must therefore have an occurrence of only one per set. Others known as MEMBER record types can have any number of occurrences. From time to time, the manager will refer to the various contents of each record.

Details such as account number, policy number, premium amount and vehicle type will probably be DATA ITEMS; other details such as dates, addresses and claim numbers may be DATA AGGREGATES. The difference is that a data item cannot be further subdivided, while a data aggregate consists of a number of data items, but can be referred to as a whole. For example, *date* has three data items, namely, DAY, MONTH & YEAR, and *address* has STREET, TOWN, COUNTRY, etc. The distinction between these two units must be made for two situations. Firstly, where a larger unit of data is always used in a program, but only a part of it may be updated, eg, in date, all three data items may be moved in a program, but only one changed; secondly, where a repeating group is involved, eg, a premium received may consist of data items policy number and amount, but may exist several times for one policy.

Each occurrence of a data item will have a specific value (for example the value of 'premium amount' may be '£91 25' or the value of 'vehicle type' may be 'saloon car'); these values will be represented in a suitable code.

These logical data units can be transferred fairly easily from the clerical example to one where programs, DBMS, and direct access storage replace the manager, filing clerk, and filing cabinets. Programs may request an occurrence of a SET by providing either a primary or a secondary key (the primary key of a member record). A number of RECORDS will then be retrieved for the program, which will manipulate the DATA ITEMS and DATA AGGREGATES stored therein.

Physical database units

Input to and output from the central processor in a computer system is accomplished by a command which causes a secondary storage device to be read from one indicator to another, and the data thus found transferred to main memory. In the case of a magnetic tape, the indicators will be inter-block gaps; in the case of magnetic disks or drums, the indicators will be address markers.

All the data recorded between these indicators form a physical record sometimes known as a block or sector. A PAGE may be one or more of these (contiguous) physical records. In practice, it is unlikely that a page will be longer than a track of a magnetic disk, and it should be possible to perform a multiple physical record read in a short time to retrieve a page. Where it is possible to vary the size of physical records in a computer system, positive performance advantages will be gained by making the page size exactly equal to the physical record size. A page can contain a number of records, and a record can span more than one page. There may be different record types to a page because there is no restriction on record/page relationships.

An AREA is simply a collection of pages, normally contiguous. There is no need for an area to be restricted to disk pack, cylinder or track boundaries, as an area will consist of a multiple of pages. The specification of areas provides a means of physically segmenting secondary storage; the page is generally that part of an area which is read into main storage at every input or output command.

It is the job of the DBMS to match requests from programs for set or record occurrences to area and page identifiers. (This is the CODASYL approach, other approaches are different in detail.) The database approach enables the introduction of the concepts of VIRTUAL data and TRANSPARENT data: VIRTUAL data is the term given to data item types whose values are calculated from other stored data items and presented to a program only when they are requested by that program. TRANSPARENT data is the term given to data items whose values are for the benefit of the DBMS, and are never seen by the programs.

Terminology cross-references

While an explanation of the CODASYL terms is useful in conveying concepts, other types of DBMS use different terms, and occasionally different concepts. Here is a list of other terms used and their CODASYL equivalence:

ELEMENT
FIELD
DATA ELEMENT
DATA FIELD
ELEMENTARY ITEM
} DATA ITEM

GROUP ITEM
REPEATING GROUP
VECTOR
SEGMENT
} DATA AGGREGATE

SEGMENT
ENTRY
} RECORD

FILE .... approx. } ALL OCCURRENCES OF A RECORD TYPE; or
DATA SET ...... approx. } ALL OCCURRENCES OF A SET TYPE.

These are all logical data units; physical data units are so widely variable according to manufacturer that such equivalence is almost impossible. All terms which refer to large areas of secondary storage, such as, EXTENT, GRANULE, and in some cases POSITION and CELL, may have some equivalence to AREA. Terms which refer to smaller elements, such as BUCKET, SECTOR, BLOCK and CELL (again) may be equivalent to PAGE. Because of wide differences in physical data notation, equivalence between terms in this area will be useful only in comprehending a move from one manufacturer to another, and should be of concern only to those who will be responsible for physical data storage.

The Schema

Once a database system has been designed, it will be possible to identify each type of data item, data aggregate, record and set, by a name or code. It will be possible to state which data item types go together to make data aggregate types and record types, and to identify which record types are members and owners of set types. A coded set of tables describing this information and stored in the computer system on direct access devices is called a SCHEMA. It is a description of the data structure which is separate from the data itself.

In the CODASYL recommendations, the Schema also describes the areas, their identifiers and page sizes, and indicates how these are related to the records and sets. In other systems, a different set of tables is used for this.

The Schema, therefore, is the view of the data, the overall logical data structure, which is held by the DBMS. Each time a program requires data, the DBMS will look up in the Schema for the details of the structure of the data requested; for example if the program requires an occurrence of a set, the DBMS will look up in the Schema which record types are required, how to find the relevant records given a certain key by the program, and perhaps also which areas the pages containing the relevant data are stored in.

The term schema is used by CODASYL, but any collection of data that describes logical data structures may be seen as a type of schema. The view of the data as seen by the program may be different from the schema which is used only by the DBMS. The program will only know of the existence of certain records and data items, and these may be in a sequence different to the sequence of storage or DBMS reference. There may be other differences between the data used by the program and the data as seen by the DBMS.

The program will need its own description of the logical data structures as well, which can be achieved by a preamble to the program itself, such as the DATA DIVISION in a COBOL program.

In a database system, it is not always possible to allow programmers to write the data division of their choice for reasons of security or control. It is more usual to provide the programmer with a standard description of the logical data to be used in a particular application. All references to data within the program will be to this description, which is called a SUBSCHEMA by CODASYL and is similar to the SCHEMA in structure.

The DBMS has the job of matching data requests on a subschema and data requests based on the schema (fig. B1 appendix B). In addition to this, the DBMS must match schema-based requests to physical data requests, based on the physical storage structure.

PHYSICAL DATA ORGANISATION

Database systems employ techniques for organising data which are designed to provide different logical record sequences from a single physical sequence.

There are three basic means of superimposing a different logical sequence of records on an existing physical sequence:

- a partitioned file;
- a chain file;
- an inverted file.

Each technique has its own variations, and each method of representation is applicable in specific environments.

A partitioned file

A partitioned file is made up of subfiles called members, and each member file is referenced by a directory which may have a number of levels according to the speed of retrieval required. Within each member subfile, the records are held in sequence, and this structure is sometimes called a physically contiguous file. An example of use of a partitioned file can be seen in figure 11.1, which shows part of a partitioned file of stock records for an electrical equipment company. The showroom directory is first searched for the required showroom (say Manchester), whose link field points to page 7 in the Maker directory. A search is then made from page 7 onwards to the required make (say, LECTRO), which is found at page 8 of the Maker directory.

It may be desirable to keep a check in each directory of the number of pages to be searched in the next directory. For example, in the Maker

Showroom Directory

Page	Key	Data	Link
1	Birmingham	–	1 (3)
2	London	–	4 (3)
3	Manchester	–	7 (4)

Maker Directory

Page	Key	Data	Link
1	ELEK	–	1
2	LECTRO	–	4
3	WUNDA	–	7
4	ELEK	–	9
5	FLEX	–	11
6	PLUGG	–	12
7	ELEK	–	14
8	LECTRO	–	16
9	PLUGG	–	18
10	WUNDA	–	20

Stock File

Page	Key (stock no.)	Data
1	01049	1 kw fire
2	01128	2 kw fire
3	01305	4 kw fire
4	00409	7 cu. ft. refrigerator
5	00410	13 cu. ft. refrigerator
6	00507	13 cu. ft. freezer
7	06042	hair drier
8	06043	hair drier
9	01049	1 kw fire
10	01050	1.5 kw fire
11	02292	13 amp plug
12	03494	flat iron
13	03495	steam iron
14	01501	fan heater
15	01510	fan heater
16	00409	7 cu. ft. refrigerator
17	00410	13 cu. ft. refrigerator
18	03490	wall heater
19	03495	steam iron
20	06100	heated roller

Figure 11.1 A partitioned file example

directory, pages 4, 5, 6 refer to the London showroom, and so the number 3 could be included in the Showroom directory alongside the link 4. Thus enabling the program to stop searching when it reaches page 7 in the Maker directory if a maker entry has not been found before then. In this case the link from Manchester would include the number 4 which is the number of pages in the Maker directory relevant to Manchester, ie, 7–10. Having found LECTRO on page 8 of the Maker directory, it points to page 16 of the stock file, which is the first page of a sequence of LECTRO equipment records at the Manchester showroom. The stock file contains member subfiles for each showroom.

This technique is often used for physical storage of logical data in a tree structure. The example in figure 11.2 is the tree structure for figure 11.1.

Partitioned files may be created without directories; these are often called sequential hierarchies. Here, two or more record types are used and partitioned so that those records which are logically retrieved together are stored serially. An example of this is shown in figure 11.3. Here customer, order, and order item records are stored so that each customer record is followed by the first order record, which is followed by the first order item record and so on. A physical structure such as figure 11.3 does not permit

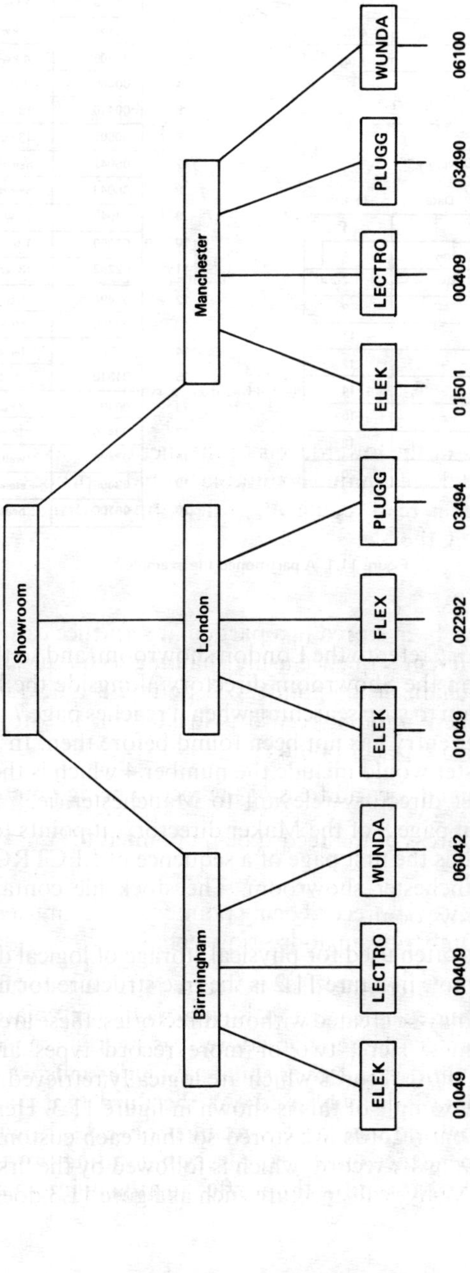

Figure 11.2 Tree structure of data in the partitioned file

Customer 1
Order 1/1
Order Item 1/1/1
Order Item 1/1/2
Order Item 1/1/3
Order Item 1/1/4
Order 1/2
Order Item 1/2/1
Order Item 1/2/2
Order Item 1/2/3
Order 1/3
Order Item 1/3/1
Order Item 1/3/2
Customer 2
Order 2/1
Order Item 2/1/1
Order Item 2/1/2
Order Item 2/1/3
Order 2/2
Order Item 2/2/1
Order Item 2/2/2
etc.

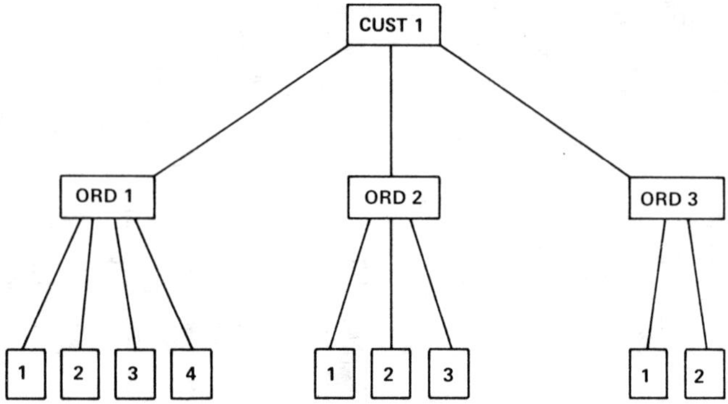

Figure 11.3 Sequential hierarchical structure

the rapid searching of the lower levels of the hierarchy, as a scan of all the records is needed. This technique is suitable for batch processing situations where the entire file is read regularly, as deletions and insertions are best handled by rewriting the file.

Chain files

Records which have been stored in a particular sequence can be processed in a completely different sequence using chaining techniques. Records are linked together in chains using pointers. A pointer is a field in one record which indicates where another record is located on the storage devices. There are three main types of pointer:

- the direct access device address of the record being pointed to;
- the relative address of the record being pointed to; this is usually the page number;
- the primary key of the record being pointed to; this must be converted into a page number by an addressing technique.

The first type of pointer, the disk address, is fastest in accessing, but provides no data independence and is very poor for volatile files. The second type of pointer, the relative address, is nearly as fast as the disk address, is usually shorter and gives some data independence, particularly when paging of data is heavily used. As the third type of pointer, the primary key, is slow and may need a number of seeks, it is not useful in chains. Data independence, however, is very high using this technique, as records can be moved around without affecting the pointer value.

The following examples which illustrate chaining use the second technique, the relative address or page number. Suppose an order file which is in order number sequence is required to be read in customer sequence, then the new sequence can be effected by using a directory of customers and a set of pointers (fig. 11.4).

Customer Directory

Page	Cust	Data	Ptr
1	A		2
2	B		3
3	C		1
4	D		–

Order File

Page	Ord	Data	Ptr	
1	001		5	
2	002		4	
3	003		8	
4	005		6	
5	009		7	
6	010		10	
7	012		9	
8	014		11	
9	017		12	
10	021		13	
11	025		16	
12	026		14	
13	029		15	
14	030		17	
15	032		–	(1)
16	034		–	(2)
17	035		–	(3)

Figure 11.4 Chaining-relative address pointers

In the Customer Directory the record for customer A has a pointer to page 2 in the order file, this page contains a record for order 002 and a pointer to page 4 which contains a record for order 005 and a pointer for page 6 and so on. This is the chain of order records belonging to customer A; the end of the chain contains a character denoting the last record on the chain. This particular format is called a 'list file', and with this type of file the programmer is unable to determine which customer a particular order is associated with if this information is not contained as part of the data in the order record, as there is no pointer to the customer records from the order record.

If a pointer to the head of the chain is inserted instead of an end-of-chain character in the last record of the chain, we have a 'ring structure'. This type of file enables the program to progress down the ring until the customer record type is reached, thus determining the customer for a set of orders. The page numbers in the records at the end of the chain must be page numbers of the directory. In the case of figure 11.4, order 032 would point to page 1 of the customer directory, order 034 to page 2 and order 035 to page 3.

Each ring in figure 11.4 is sequenced in ascending order number. This will be particularly useful when writing a program to produce invoices, statements or other such items in sequence, since the ring is followed as it appears for each customer until a pointer to the original customer record is reached. A sequenced chain means that each new order as it arrives must be

Page	Cust	Data	Ptr
C1	A		2
C2	B		3
C3	C		1
C4	D		–

Customer directory (PTR1)

Page	Ord	Data	Ptr 1	Ptr2
1	001		5	10
2	002		4	4
3	003		8	6
4	005		6	9
5	009		7	7
6	010		10	13
7	012		9	12
8	014		11	16
9	017		12	14
10	021		13	3
11	025		16	8
12	026		14	17
13	029		15	S2
14	030		17	S4
15	032		C1	11
16	034		C2	S3
17	035		C3	S1

Order file

Page	S'man	Data	Ptr
S1	W		5
S2	X		1
S3	Y		15
S4	Z		2

Salesman directory (PTR2)

Figure 11.5 Records participating in two chains

added to the ring in a particular position. This will involve reading the ring until the order with a primary key higher but nearest to the insertion is found. The previous record is then modified to point to the new record, and the new record points to the next record, which may prove a lengthy process. In an unsequenced ring, records may be added in the first position, which will be faster. Also, performance may be optimised by placing the most-used records on the ring close together, avoiding lengthy seeks.

A record may participate in more than one chain, eg, orders may be linked according to salesman as well as customer for sales analysis (fig. 11.5). In this example, a search of the order file from customer A yields order 002, 005, 010, 021, 029 and 032; a search from salesman Y yields orders 032, 025, 014 and 034 in that sequence. On the physical sequence of order numbers are imposed one logical sequence and one logical grouping: customers being the sequence and salesmen being the grouping.

Now what happens when order 025 is deleted? There are two methods available; firstly, instead of deleting the records, markers are placed to indicate deletions.

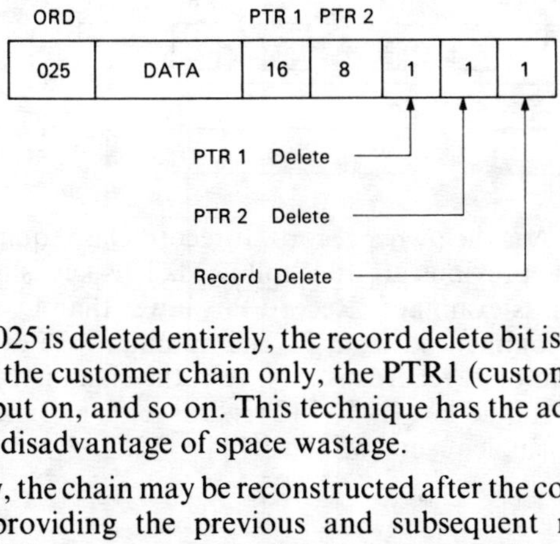

When order 025 is deleted entirely, the record delete bit is put on. When it is deleted from the customer chain only, the PTR1 (customer pointer field 1) delete bit is put on, and so on. This technique has the advantage of fail-safety, and the disadvantage of space wastage.

Alternatively, the chain may be reconstructed after the complete removal of a record, providing the previous and subsequent records can be identified. For this purpose, a second pointer is used, this time pointing in the reverse direction; in this way, the previous records in each chain can be altered to point to the subsequent records, thus omitting completely the deleted record which may then be physically removed.

While this technique saves storage *after* deletion, storage *before* deletion is greater for accommodation of the second pointer, and the deletion process takes longer. Even if 'prior' pointers are not used, record addition time is improved in cases where additions are at the end of the chains when the record at the head of the chain contains a pointer to the record at the end of the chain.

When there is a requirement to establish the identity of the record at the head of the chain and this information is not stored in the data of the record in the chain, then a pointer to the head of the chain can be inserted into each record. This pointer may either be the page number of the 'owner' record or the primary key; the primary key providing more data independence.

Sequenced chains may be organised so that searches take the minimum time. There are two basic techniques: skip searching, and using a multilist.

Skip searching

Skip searching involves dividing the records into groups; the page number of the lowest key record in the group before the head of the chain is the pointer in the owner record. The record primary key is tested, and if still too high then the pointer to the lowest record in the next lower group is followed, this being held in the record under examination, and so on. In this

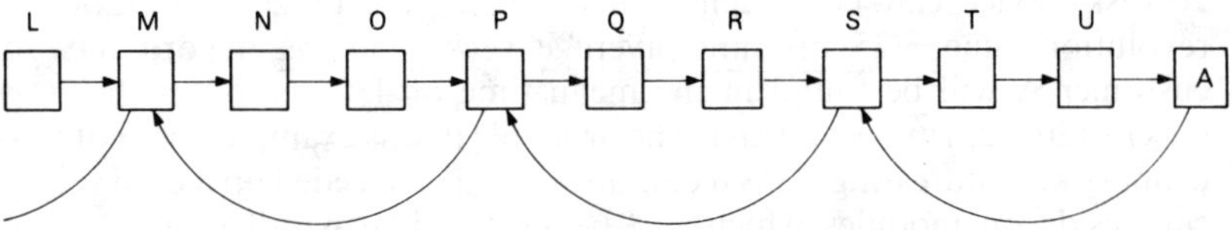

example, record A is the owner record. If record Q is required, the pointer to the start of the previous group is followed. This key is too high, so the next group down is examined. Record P is lower than Q, so a sequential search for Q is commenced from P. The optimum size of each group of records in a chain of N records is $\sqrt{N}$, so if the chain is 144 records long then each group should be $\sqrt{144} = 12$ records long, and the mean number of searches for a single record will be $\sqrt{144} = 12$.

Multilist

In a *multilist*, a sequenced chain is divided into sections. A directory gives a list of the beginning page number of each section, thus the mean searches are reduced to:

$$\frac{\frac{Nc}{Ns} + 1}{2}$$

where N_c is the number of records in the chain, and N_s the number of sections. This may be speeded even further if each section resides on a cylinder of a moving head disk, as an entire section can be read without moving the disk arm. As an example, a chain of 1048 records is sequenced and split into sections below.

CUST	Key	Cy1	Key	Cy1	Key	Cy1	Key	Cy1
A	050	102	090	107	170	109	210	250
B								
C								

This shows the customer directory for a multilist order file. Supposing the record required is order 098 for customer A, then an examination of the directory yields cylinder 107 as the section to search. An entire cylinder on a 20 disk device can be searched in a mean time of 20 revolutions; if revolution time = 25 ms, and average seek = 30 ms, record 098 for customer A will be found in the mean time of 430 ms (30 ms (average seek) + (20 × 25) ms = 430 ms). The order file in this example does not have primary key addressing. This technique is further speeded up if each section can reside on modules which can be searched in parallel, eg, different devices.

Chaining techniques may be used to represent a wide range of data structures. The main limitations are the search times and the duration of insertions and deletions, both of which can be minimised using the techniques described.

Inverted files

Inverted files are organised so that records may be retrieved by giving values of any data items to find a record or group of records. It may be necessary, for example, to retrieve from an employee file all records of employees with salaries less than £3,000 and who work in the Sales Department. This may be done by searching each record in the file and retaining only those which meet the specifications, or by accessing only those records with the correct specifications by means of an inverted file. A complete search of the file can only be useful in a batch system, where there is no time limit on the response, as file scans can be lengthy.

There are four basic methods of organising inverted files:

- secondary indexing;
- partial inversion;

- full inversion;
- bit map indexing.

Secondary indexing

The records are held in primary key sequence, and secondary indexes are maintained giving the value of secondary keys. For example, a product file is maintained in product number order, as this is the primary key. Suppose it is necessary to retrieve products of particular colours, then a secondary index of the secondary key 'colour' may be maintained (fig. 11.6).

Colour	Products				
red	001	003	007	009	
blue	002	004	010	011	016
green	005	006	008		
white	012	013	014	015	

'Colour' secondary index

Product	Data
001	
002	
003	
004	
005	
006	
007	
008	
009	
010	
011	
012	
013	
014	
015	
016	

Product file

Figure 11.6 Secondary indexing example

Each entry in the secondary index is for a value of the secondary key 'colour'. Within each entry is a list of primary keys of records with the particular value for colour, eg the entry for 'Red' in the colour secondary index contains a list of product numbers which are the primary keys for products 001, 003, 007, and 009. This form of index, where the pointers are all within each entry in the index is called a 'POINTER ARRAY'. For one particular file, there may be as many secondary indexes as there are fields in the record, although in practice only those fields selected as secondary keys need be considered for indexing.

Another method of secondary indexing can be used when records are held in any sequence and are not indexed on primary key. In this case records can only be retrieved by the secondary indexes, which use page numbers (ie, relative addresses) to point to the records. Figure 11.7 shows the same product file with two secondary indexes and without use of primary keys.

Colour	Pages				
red	6	16	3	11	
blue	1	2	12	5	10
green	15	4	7		
white	14	13	9	8	

Colour Index

Material	Pages			
plastic	5	7	12	
metal	15	14	1	3
wood	11	9	6	
fabric	13	16	4	
fibre	2	8	10	

Material Index

Page	Product	Data
1	002	
2	004	
3	007	
4	006	
5	011	
6	001	
7	008	
8	015	
9	014	
10	016	
11	009	
12	010	
13	013	
14	012	
15	005	
16	003	

Figure 11.7 Two secondary indexes

Again, pointer arrays are used in the indexes, and for each search argument that will be used there must be an index. Retrieval of records once an index entry is found is quicker when using relative addresses rather than primary keys; when primary keys are used the page number must be calculated. Insertion of new records is quicker using this method as the location of the record does not depend on any sequence, and a new record can be placed at the next available location. However, when page numbers are used in index entries rather than primary keys, any time the file is moved from one position to another, for example at file maintenance time, then the page numbers in the indexes need to be updated because they will have changed.

Customer index

Page	Key	Data	Pointer
C1	A	–	1
C2	"	–	10
C3	"	–	15
C4	"	–	19
C5	B	–	4
C6	"	–	14
C7	C	–	5

or

A	I	10	15	19
B	4	14		
C	5			

Order file

Page	Key	Data	Pointer
1	001	–	3
2	002	–	–
3	004	–	7
4	005	–	8
5	007	–	6
6	009	–	9
7	010	–	–
8	011	–	12
9	013	–	–
10	017	–	11
11	021	–	13
12	027	–	–
13	030	–	–
14	032	–	16
15	039	–	17
16	044	–	20
17	045	–	18
18	047	–	–
19	051	–	–
20	059	–	–

Figure 11.8 Partially inverted file

Partial inversion

The pointer array is used in a partially inverted file as an indexing technique. Selected attributes are stored in indexes, and each entry in the index is a pointer to part of the main data.

With partial inversion, however, chaining of records is used although the main feature of these files is that the length of the chain is limited. Figure 11.8 shows a partially inverted file with the chain length limited to three records. Customer A has 10 orders and, as each chain length is limited to three records, there are four chains linking these orders to the index entries for customer A. The first pointer associated with customer A in the index refers to page 1 which contains the record for order 001. The pointer field of order 001 points to page 3 which contains the record for 004 and this points to page 7 which is the end of this chain. Three more chains, link orders to customer A, and the head of each of these is held as a list of addresses in the index. Partial inversion considerably speeds up the processing of chains because each chain is restricted to a more manageable size. This technique is really a variation of the multilist technique described earlier.

Index maintenance is a major activity in secondary index and partial inversion systems. Each time a record is updated, then all relevant indexes must also be updated. With the pointer array organisation, difficulties may be encountered due to excessive variation in length of entries. For example, if in one product file 75 % of the products were red and the file had 10,000 records, then the index entry for colour red would contain 7,500 pointers and would be by far the biggest entry. This situation may be alleviated by using chain file techniques instead of pointer arrays, although these may be unsuitable on grounds of search time.

Full file inversion

A fully inverted file contains an index for each attribute of data held, each field in the record being a potential search key. Figure 11.9 shows one type of fully inverted file in which there are four main parts:

- an attributes index;
- a values index;
- an occurrence index;
- data values.

The ATTRIBUTES index, lists all field names under which data is held. Suppose that we are constructing a fully inverted file of records of the following format:

PART No.	CLASS	TYPE	COLOUR

Attribute	Page	Qty
part	1	6
class	7	3
type	10	4
colour	14	2

Attribute Index

Page	Value	Page	Qty
1	001	101	1
2	002	102	1
3	003	103	1
4	004	104	1
5	005	105	1
6	006	106	1
7	A	107	3
8	B	110	2
9	C	112	1
10	A1	113	1
11	A2	114	1
12	B1	115	2
13	C2	117	2
14	RED	119	4
15	BLUE	123	2

Values Index

Page	Part	Page
101	001	150
102	002	151
103	003	152
104	004	153
105	005	154
106	006	155
107	001	150
108	002	151
109	003	152
110	005	154
111	006	155
112	004	153
113	002	151
114	003	152
115	005	154
116	006	155
117	001	150
118	004	153
119	002	151
120	003	152
121	004	153
122	005	154
123	001	150
124	006	155

Occurrence Index

Figure 11.9 Fully inverted file

This would result in four entries in the attribute index, one for each field in the record. Alongside each entry is a pointer to the appropriate part of the values index, where a list of values for that attribute is found.

It may be desirable to include a quantity of entries in the values index with the pointer from the attribute index to facilitate the search.

The VALUES index contains an entry for each value for a specific attribute; for example, there is one entry for the colour RED. With each entry is a pointer (page number) into the occurrence index, as well as the quantity of associated entries in the occurrence index. This inversion is now complete with the exception of the data values, which are the records themselves.

The OCCURRENCE index is used to link a particular attribute value to a record. For example, let us search the inverted file for parts which are both RED and type A1. First, the attribute index is searched for attributes COLOUR and TYPE. This leads to page 14 in the values index, where 2 pages are read and page 10 where 4 pages are read. Thus pages 10, 11, 12, 13, 14 and 15 from the values index are searched for colour RED and type A1. These are found at pages 10 and 14, which indicate that 1 entry at page 113 and 4 entries from page 119 of the occurrence index must be searched.

The search now is for part numbers which appear on both lists in the occurrence index, as these will possess the required attribute values. A comparison of the two lists can be made to produce part number records with the same type (A1) and the same colour (RED). Part 002 appears on both lists, and the occurrence index indicates that this is stored at page 151 in the data file.

Thus a fully inverted file may consist of a set of three tables referring to the data records. The values index and occurrence index may, in a large file, be long and require lengthy searches at accessing times. These searches may be minimised using index sequential, binary searching or skip searching techniques.

Bit map indexing

In bit map indexing, an inverted index is maintained with an entry for each attribute/value pair. A bit position in each is reserved for each record in the file, and if the record contains the attribute/value pair, then the corresponding bit is 1, otherwise it is 0. The inverted bit index for figure 11.9 would be as figure 11.10.

There is the same number of bits against each attribute/value pair as there are records in the file – in this case, six. Logical operations on an inverted bit index are straightforward and fast in execution. To find records with colour RED and type A1, the appropriate entries are found in the index, and the bit strings (or sequences) associated with them are compared

	001	002	003	004	005	006	Part no.
Class A	1	1	1	0	0	0	
Class B	0	0	0	0	1	1	
Class C	0	0	0	1	0	0	
Type A1	0	1	0	0	0	0	
Type A2	0	0	1	0	1	0	
Type B1	0	0	0	0	1	1	
Type C2	1	0	0	1	0	0	
Colour red	0	1	1	1	1	0	
Colour blue	1	0	0	0	0	1	

Figure 11.10 Bit map indexing

(ie, logically ANDed); where two 'one' bits exist, a 'one' bit is the result. In all other cases, a zero bit is the result. Using the example:

$$\begin{array}{rr} \text{AND} & \begin{array}{r} 010000 \\ 011110 \end{array} \\ \hline & 010000 \end{array}$$

All other logical operations are facilitated, therefore a wide range of data requests can be met, eg, we may wish to know which parts have class A OR B, AND type C OR COLOUR RED.

A limiting factor, however, is the length of the bit strings, as a file of 10,000 records would require 10,000 bit positions in a logical operation. Also, an active file would require that the index be frequently updated and modified, which represents a maintenance overhead.

DESIGNING DATABASES

Databases are normally implemented by using a package called a Data Base Management System (DBMS). Each particular DBMS has somewhat unique characteristics, and as such, general techniques for the design of databases are limited. DBMSs fall into two broad categories: 'pointer-driven' systems and 'table-driven' systems. The latter are inverted-file systems which allow the user to set up and maintain a database which may be searched using a wide range of different keys, via either a set of commands supplied with the package, or 'calls' from a host language such as COBOL or PL/1. These systems are generally straightforward to implement; the user specifies the records and fields, indicates which of the fields will be keys, and supplies these parameters to the DBMS, which will then set up the database. Pointer-driven DBMSs use techniques such as partitioning and chaining. These are normally host-language systems which use high-level language verbs coded within application programs. The design of the database using such systems will have a crucial effect on the performance and flexibility of the end result. Complex data structures may be used, and the effort needed at the design stage is much greater than with table-driven systems.

Good examples of pointer-driven DBMSs are the packages produced to the CODASYL recommendations. Techniques for designing the logical data structures for CODASYL Data Base Definition have been developed, and this section examines one in detail.

A most important step in the process of CODASYL database implementation is the design and implementation of the SCHEMA. The design features of the schema will have a significant effect on the performance and flexibility of the systems using it.

Schema design

A well-designed schema will enable the programmers to write efficient, data-independent programs which will not need major modifications whenever new data or storage techniques are added to the system.

Schema design is mainly concerned with choosing 'record' types and content, and 'set' types, content and characteristics. The data items which will go into the database are first defined, then grouped together in records, and the records grouped together in sets. The first stage is to analyse the data into third normal form (as in chapter 9) and then to use a data mapping technique to decide which record and set groupings should be made.

As an example of the methodology, the following list gives the data item types available to an employee database:

- employee no.;
- name;
- address;
- date of birth;
- national health insurance no. (NHI no.);
- department no.;
- job no.;

- grade;
- salary;
- bank no.;
- bank account no.;
- skill detail;
- experience detail;
- education detail.

In order to decide which record types should be made, it is first necessary to identify the RETRIEVAL KEYS which are to be used. In this case, EMPLOYEE NO., NAME, NHI NO., DEPARTMENT NO. and JOB NO. may be the keys which the system will make available to the programs. In practice, this choice is very important, and will entail a very thorough examination of the proposed system outlines to determine which values will be available to act as keys. RECORD TYPES may contain the following data item types:

- the primary key: unique to the record, used to store the record at a disk address. (When page addressing techniques are used, this data item value may not be unique to a record.);
- data items: which are not used as secondary keys, and which are not repeating groups;
- secondary key data items: which are unique to the record, and may, if fixed length and a suitable code for addressing, be stored in the record rather than as a set;
- repeating group data item or data aggregates: which have a limited and fairly small occurrence per record, and which do not have separately-stored secondary keys.

SETS may relate:

- records to repeating groups which have a variable and large number of occurrences per record, or which themselves have separately stored secondary keys;
- secondary keys to records where the secondary key is not unique to the record, but common to many record occurrences (the one-to-many relationship).

The logical design work must have produced system outlines before the database can be properly designed. It is not important, at this stage, to distinguish between *primary* and *secondary* keys, this distinction will come later. The keys will be the values available to the program to retrieve data, and these may be determined by the general specifications of the functions of the various application areas. For example, a system which is adding customer orders to the database will have available, from input, the CUSTOMER NO. and the ORDER NO., these will be the KEYS for this function. Once the keys have been determined, the relationship between each key item and each relevant and related data item must be examined, perhaps in a matrix like the one in figure 11.11.

Key Item	Data Item	Is Data Item a Key?	One-to-one One-to-many Many-to-one Many-to-many	Length of Data Item	F or V	Occurrences of Data Item	R or S
empl. no.	address	no	one-to-one	0–200	V	1 per empl.	R
empl. no.	dept. no.	yes	many-to-one	8	F	30	S

Figure 11.11 Specifying and analysing data items

The key data item name is entered in 'key item' column, and each related and relevant data item (including other keys) are entered one at a time in the adjacent column (the 'Data Item' column). Then test to see whether the second data item is a key and indicate this in the third column. Next, the relationship between the key data item in the first column and the data item in the second column must be examined; it will be either one-to-one, one-to-many, many-to-one or many-to-many. In the example, the key data item is EMPLOYEE NO. and the first Data Item is ADDRESS; this is not used as a key (NO. entered in column 3) and it is in almost every case a one-to-one relationship. The second data item to be related to EMPLOYEE NO. is DEPT. NO.; this is used as a key (YES entered in column 3) and the relationship between EMPLOYEE NO. and DEPT NO. is many-to-one. The next three columns in the matrix indicate the length of the data item, whether the length is fixed or variable, and the number of occurrences of the data item; these values will help decide whether or not to store a repeating group in a record.

Once the matrix is complete, the final column can be used to denote whether or not a particular relationship is to be stored as a record type or set type (R or S). The table in figure 11.12 can be used to help decide the contents of the final column.

		One-to-one	One-to-many	Many-to-one	Many-to-many
Is Data Item a Key?	no	record	record or set (1)	record or set (2)	set
	yes	record or set (3)	set	set	set

(1) Record if occurrences are fairly limited in number and fixed length

Set if occurrences are large in number and variable, or if data item has another separately stored secondary key

(2) Record if the data item is short to medium in length

Set if the data item is very large

(3) Record if the data item is fairly short, fixed length and of a suitable code to provide an adequate addressing technique

Set if the data item is long, or variable length, and unsuitable to be used as the basis of an addressing code.

Figure 11.12 Storing relationships between data items in records or sets

Key Item	Data Item	Data Item = Key?	One-to-one One-to-many Many-to-one Many-to-many	Length of Data Item	F or V	Occurrence of Data Item	R or S
empl. no.	name	yes	many-to-one	32	F	n	S
″	address	no	one-to-one	0–200	V	1 per empl.	R
″	NHI no.	yes	one-to-one	10	F	″	R
″	date of birth	no	many-to-one	6	F	″	R
″	dept. no.	yes	many-to-one	8	F	n	S
″	job no.	yes	many-to-one	8	F	n	S
″	grade	no	many-to-one	8	F	n	R
″	salary	no	many-to-one	8	F	n	R
″	bank no.	no	many-to-one	6	F	n	R
″	bank acc. no.	no	many-to-one	8	F	1 per empl.	R
″	skill detail	no	one-to-one	0–200	V	n	S
″	experience	no	one-to-many	0–200	V	n	S
″	education	no	one-to-many	100	F	<20 per empl	R
skill type	skill detail	no	one-to-many	0–200	V	n	S

Figure 11.13 Full matrix for example

The next stage is to select the Key Items which are PRIMARY KEYS. A primary key should be unique to a record (in exceptional cases where certain page addressing techniques are used, the primary key may not be unique). A primary key should be suitable for use as a parameter for storing the record, fixed length, reasonable size. The primary keys will also be values which are available to a program when it is run, these may be values from input documents, or input or master data. Once a primary key has been established, all other data items which are related and have R in the final columns, become data item names in the record for which the selected key item is the primary key. This may be repeated for all primary keys until a list of record types, their primary keys and other contents have been established. The remaining entries in the matrix will be represented as sets.

Using the example in figure 11.13 as an illustration of this technique, the first primary key is employee no. and relates to the following data items having R in the final columns:

- address;
- NHI no.;
- date of birth;
- grade;
- salary;
- bank no.;
- bank acc. no.;
- education.

These can be grouped as a record with the primary key.

All the other relationships involving Employee No. will be represented as sets, namely:

- name;
- dept. no.;
- job no.;
- skill detail;
- experience.

The other primary key will be skill type, which is related to Skill Detail by a set. Thus, the following record types and set types are shown in the schema:

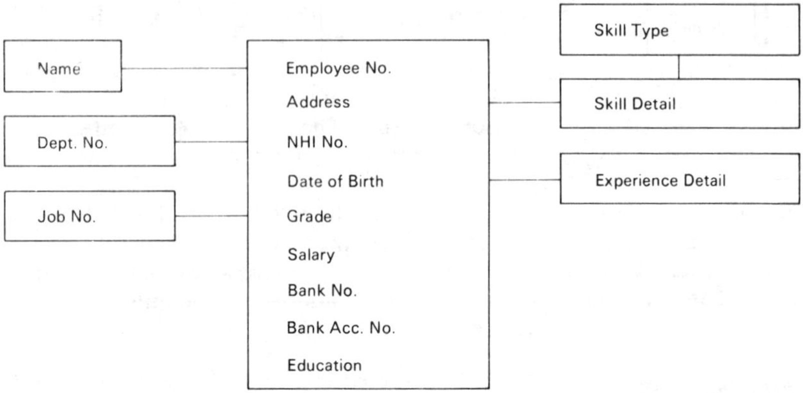

This diagram will form the basis of a schema. The complete documentation of a schema will include a number of indications showing primary keys and set characteristics. To complete this preliminary diagram, information regarding the basic characteristics of the sets must be shown; the data structure diagram conventions in chapter 9 can be used:—

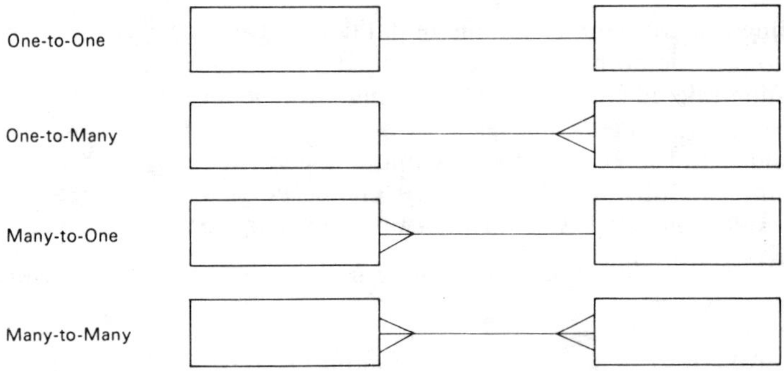

Using these conventions the schema can be redrawn showing the basic characteristics of the sets, as facing page.

Physical organisation of data

The various techniques of data organisation in databases have been described earlier, and can now be examined for appropriateness to particular circumstances.

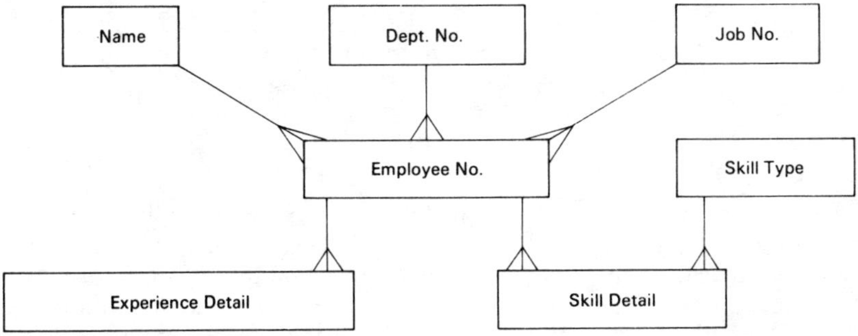

Partitioning

A partitioned file may be used where a large file is to be reduced to sub-files, and where selective accessing of subfiles is required (eg, to examine the total sales for one region from a file which contains others). By partitioning the file, a direct access may be made to the subfile of orders for the region in question. The partitioned file may be held as a sequential hierarchy if direct access to each subfile is not required. A sequential hierarchy will only be of use in a batch processing environment.

Chaining

Chaining is a common technique in databases. It may be used in several ways to provide both flexibility and speed of access. In general, the only restriction offered by the use of chaining is the length of search down a chain of records. If real-time response is required, then either chaining may be avoided and some inventory technique used instead, or the use of skip searching, multilisting or sequenced chains employed. It must be remembered in this case that sorting chains into sequence can be time- and resource-consuming.

Inversion

Each of the four major inversion techniques is appropriate to particular circumstances. Secondary indexing is used where the file is regularly processed in primary key sequence, the required retrieval time is not critical and the file is frequently maintained. Partial inversion is more appropriate where the file is not processed in primary key sequence, the required retrieval time is fairly fast, the file is active, and the maintenance requirement is low. Full file inversion is appropriate when the search criteria are highly variable and the required retrieval time is fast. Updating a fully inverted file is, however, a lengthy process, and in real-time systems, fully inverted files may be subject to deferred updating in order to be able to satisfy response time requirements. If the file is not too large for a bit array of the data to be accommodated in main store, then bit map searching gives good results where search criteria are mixed and variable.

SUMMARY

The database approach to system design places great emphasis on the integration, integrity and independence of data; this involves the separation of logical and physical data structures so that the physical storage of the data can be altered without affecting the logical structure, and *vice-versa*. Physical data organisation in databases includes the methods of file organisation and access described in the last chapter but also some additional techniques for providing a variety of logical structures from a single physical structure; these techniques include partitioning, chaining and inversion. The selection of appropriate techniques is often constrained by the database management software available, but the design of databases always requires careful data analysis and schema design.

12 Output and Input Design

INTRODUCTION

As a result of the logical system design activities the systems analyst will have defined the nature of the output reports required by the system and the input and stored data required to produce these outputs. The physical design activity takes these outlines to a detailed level and determines the media to be used to produce the output and to capture the input data.

This must begin therefore with an appraisal of the earlier documentation and must be done alongside, and as an integral part of, file and computer procedure design. The specification of user requirements is the starting point for the appraisal and the detailed physical design must be done in the light of this and with continuing user involvement.

The normal procedure is to design the outputs in detail first and then to work back to the inputs. The outputs can be in the form of operational documents, lengthy reports replies to enquiries or summarising graphs. The inputs can consist of batched records or individual real-time messages generated at terminals. The input records have to be validated, edited, organised and accepted by the system before being processed to produce the outputs. Both output and input have to be specified in detail in terms of computer procedures and clerical handling. The systems analyst has to define the method of capturing data and the input program and the format of the output and its use by the user depatment. This chapter examines output and input design and specification.

OUTPUT DESIGN

Outputs from computer systems are required primarily to communicate the results of processing to users (or sometimes to other systems, including machine-based systems). They are also used to provide a permanent ('hard') copy of these results for later consultation. There are various types of output required by most systems; the main ones are:

- external outputs, whose destination is outside the organisation and which require special attention because they project the image of the organisation;
- internal outputs, whose destination is within the organisation and which require careful design because they are the user's main interface with the computer;
- operational outputs, whose use is purely within the computer department, eg program listings, usage statistics etc;
- interactive outputs, which involve the user in communicating directly with the computer (often called dialogues);
- turnround outputs, ie, re-entrant documents, to which data will be added before they are returned to the computer for further processing.

Output definition

The outputs may have been defined during the logical design stage. If not, they should be defined, at the beginning of the output design, in terms of:

- type of output;
- content (headings? numeric? alphanumeric? totals? etc);
- format (hard copy? screen? microfilm? etc);
- location (local? remote? transmitted? transported? etc);
- frequency (daily? weekly? hourly? etc);
- response (immediate? within a period? etc);
- volume (number of documents? growth? etc);
- sequence (account number? within sales area? etc);
- action required (bursting? error correction? etc).

The content of the outputs must now be defined in detail.

Data items

The name given to each data item should be recorded and its characteristics described clearly in a standard form:

- whether alphabetic or numeric;
- legitimate and specific range of characters, eg minimum, maximum, fixed values or ranges;

- number of characters;
- position of decimal point, arithmetical sign or other indicator.

At those installations which maintain a register of the name and description of each data item used in the installation, many of these items may already have been registered. Any data item not yet defined must be identified and recorded before output design can proceed. The objective is to prevent the same data item being referred to by various names, or the same name being used to describe different items.

Data totals

There is often a need at output to provide totals at various levels. These usually result from computer processing, and must therefore be contained on computer files. Their source must be identified, and they must be defined and registered as data items.

The systems analyst must know:

- what are the various total levels required, eg column totals, subtotals, grand totals;
- whether they should appear in a vertical column, offset or at the end of a horizontal line;
- what will cause them to occur, eg change of key, special markers or other conditions;
- how will they be identified as totals, eg implied or by special description;
- if there are brought forward or carried forward totals;
- what is their value and how should this be shown, eg numeric, money, weight, volume;
- what is their accuracy requirement, eg rounded to units of one, hundreds, thousands.

Data editing

It is not always desirable to print or display data as it is held on a computer. The systems analyst must ensure whether the form in which it is stored in the computer (files) is suitable for the output, for example:

- will decimal points need to be inserted?
- will leading zeros need to be suppressed or are they required to be shown?
- where are money symbols to be shown, to the left of the first digit, in a fixed position or replacing each leading zero?

- is the alignment, left or right justified? (Numeric items are usually right justified and alphabetic items left justified.)

If any of these have not been previously specified by the users, agreement must now be reached. The 'picture' convention can be used for specifying these items.

Other necessary editing features are feasibility checks to prevent impossible and questionable results being printed or displayed (eg negative or impossibly large values on pay slips, cheques, invoices). Standard library software routines may be used to provide this protection, but the systems analyst must be aware of these needs.

Output media

The next stage for the systems analyst is to determine the most appropriate medium for the outputs. This will involve consideration of a wide range of devices, including line printer, graph plotter, typewriter, visual display unit, magnetic media and microfilm. The choice of output medium will be affected by all kinds of considerations but the main ones will be:

- the suitability of the device to the particular application;
- the need for hard copy (and number of copies required);
- the response time required;
- the location of the users;
- the software/hardware available;
- the cost.

Once the medium has been chosen the output can be specified in detail appropriate to the device to be used.

OUTPUT SPECIFICATION

As the details are assimilated and the systems analyst reaches an understanding of the users' requirements, a mental picture of various alternative layouts will gradually emerge. Even if an outline layout has been provided at an earlier stage, this may be improved upon. Some output requirements offer less scope, eg, preprinted stationery, statutory documents, but their design must still be specified and documented; data items still need to be accurately defined and arranged for clarity and easy comprehension. The systems analyst has two specific objectives at this stage:

- to interpret and communicate the results of the computer part of a system to users in a form which they can understand and which meets their requirements;

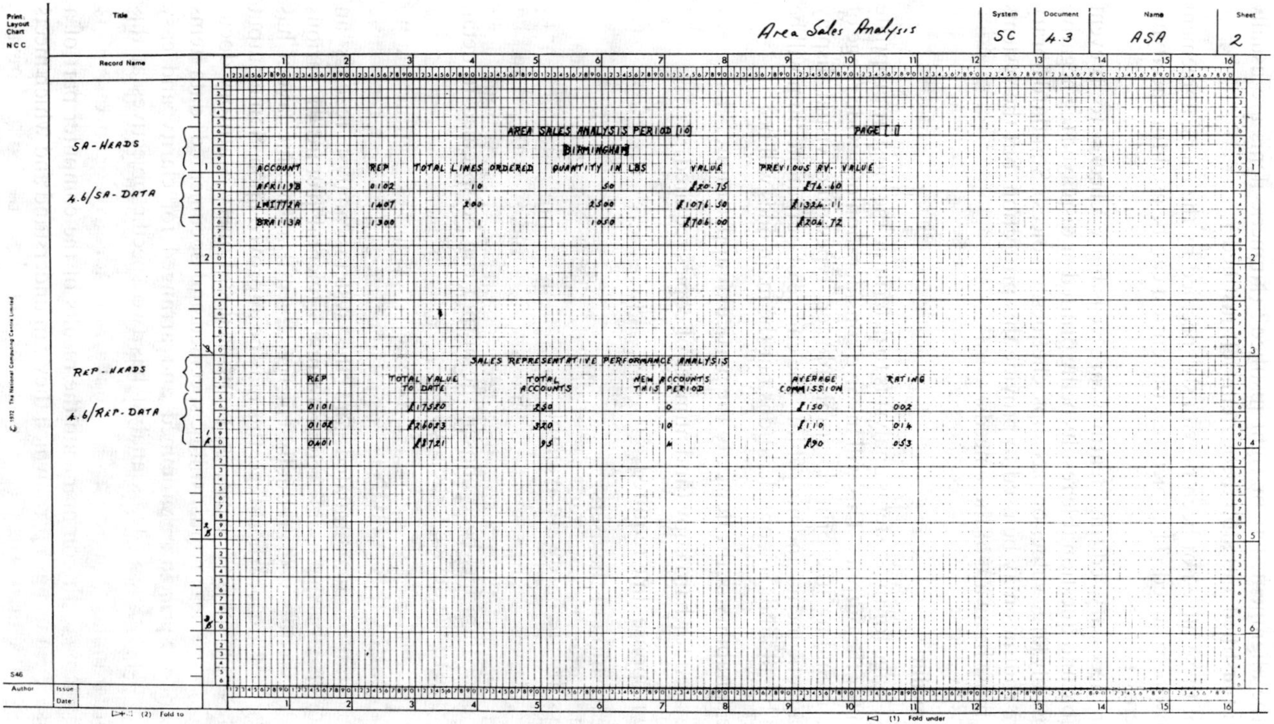

Figure 12.1 Print layout chart

- to communicate the output design specification to programmers in a way which is unambiguous, comprehensive and capable of being translated into a programming language.

Print layout charts

The layout of outputs will be normally specified on a layout chart. The form shown in figure 12.1 is a print layout chart which represents the printing positions of a line printer, usually ten character positions to the inch across the width and line spacing at six lines to the inch. Columns are headed by numerals ranging from one to the maximum number of print positions available, 160 in the example.

The proposed column headings, subheadings, some representative data items and totals may be written character by character in their appropriate position on the chart, which when complete gives a detailed representation of the output. As each square on the layout sheet corresponds to actual printer character positions it is possible to show the exact spacing of the output. The layout of two output analyses, one by account and the other by representative, is shown (fig. 12.1).

Headings consist predominantly of constants, which are fixed characters permanently in memory, and entered in their appropriate positions on the layout grid. If the headings contain any variable items, these can be indicated on the grid by enclosed brackets and their attributes shown on a list attached.

The final design layout must be approved by the user and communicated in detail to the programmer. The user's requirements are quite different to the programmer's. The user needs an example of the output as it will be received, whereas the programmer needs a precise specification which describes the output in all its foreseeable conditions and detail.

Before preparing a specification for the programmer, it is prudent to ensure that the design is acceptable to the user. At this stage although the result usually has to be simulated, it should be produced as far as possible on the actual medium and device proposed, eg, paper or screen using a line printer or visual display unit. If this is not possible, an example document, or a paper facsimile of the screen can be drawn with the output data correctly positioned. As a rule the specification for the programmer is not suitable for the user.

Specification of printed output for the programmer

The layout chart is useful for designing and communicating output design to the programmer, but by itself, it is not always adequate or effective. It is difficult to convey the various combinations of line and data formats that may arise, even using supporting narrative. There is a place for narrative, but it should be brief. Detailed information must be provided to show the

Computer Document Specification

NCC

Document title	System	Document	Name	Sheet
Area Sales Analysis	SC	4.3	ASA	1

Stationery ref.	Width	Depth	Number of parts	
Std. listing			2	Blank/pre-printed

	Average	Maximum	Growth rate
Pages	per area 10	16	2 % per annum
	total 60	80	or determining factor
Lines per page	28	28	

OUTPUT ONLY

Page and line spacing/stepping: Double line spacing

Ribbon type	Ribbon life	Printer speed	Lay-out chart ref.	Control loop Ref.
Std			ASA/2	Std list

Part No	Trim/Burst	Distribution	Line	Channel
1		Area office – Sales Manager		
2		Head office – Accounts		

MACHINE READABLE ONLY

Clear area – distance from edges	Reading method and font
	Source

Level	Record Name	Size	Unit	Format	Occurrence
6A	AREA SALES ANALYSIS				6
6B	PAGE ACCOUNTS ANALYSIS	4-28	L		1-15
C	SA-HEADS	3	L		
C	4.6/SA-DATA	1	L		1-25
6B	PAGE REP ANALYSIS	5-23	L		1
C	REP-HEADS	3	L		
C	4.6/REP-DATA	1	L		} 2-20 according to area

S 43

Author | Issue | Date

Figure 12.2 Computer document specification

layout of the output and the structure of its contents. Examples of forms which can be used and which also serve as a standard check list are the Computer Document Specification (fig. 12.2) and Record Specification (fig. 12.3).

Computer Document Specification

Both the information to be printed and the medium to be used must be described. On this form the upper part is used to describe the physical details and the lower part the logical structure of the output. (The middle part is used to specify any document which is to be read by a computer.)

This specification is cross-referenced to the layout chart (ASA in the example). The physical description of printed output consists of:

- stationery type or reference to be used, its width, depth and the number of parts;
- average and maximum number of pages, lines per page and any future increase expected;
- line spacing and page changing requirements;
- destination of each copy of the output.

The conventions used to describe the logical structure are the same as those used to describe the structure of a file in chapter 10. All the records relating to a particular computer output constitute an output file, and its structure is shown in terms of records and groups of records within the file. A record may consist of a single line or a block of print.

In any one output, a distinction can be made between a heading record and a data record; these may be identified from the sample design on the layout chart. The programmer will need to know the logical structure of the output 'file', ie, the relationship between the different types of record to be printed. As described in Chapter 10 this relationship is indicated by assigning a 'level' letter to each record. Having decided each record, any further description is useless without some form of precise identification. A unique name, by which it will always be referred, should therefore be assigned.

Since a record is always a line or block of lines, its size will refer to the number of lines or pages it occupies on the computer document. In the example, L stands for lines and P for pages.

Another factor which affects the structure and determines the size of the computer output is the number of times each record or group occurs within the next higher level. This is shown either as a fixed figure, as a minimum and maximum range or a blank which indicates an occurrence of once only.

Record Specification — NCC

Record description	System	Document	Name	Sheet
Rep. Analysis Data Line	SC	4.6	REP-DATA	1

Medium	Record length	Record format	Record size	Words / Characters / Bytes	File specification refs.	Lay-out chart ref.
Line printer	Fixed ☑ Variable ☐	Fixed ☑ Variable ☐	120	Words ☐ Characters ☑ Bytes ☐	4.3/ASA	4.3/ASA

Ref.	Position From	Position To	Level	Name In system design	Name In program	Data Type	Size	Alignment	Picture	Occurrence	Value Range
1	6	V	01	REP-DATA			109				
2	14	17	02	REP		C	4		9(4)		
3	30	35	02	VALUE TO DATE		C	6		£(5)9		
4	50	52	02	TOTAL ACCOUNTS		C	3		ZZ9		
5	69	71	02	NEW ACCOUNTS		C	3		ZZ9		
6	92	95	02	AV COMMISSION		C	4		£££9		
7	107	109	02	RATING		C	3		999		

S 44

Author | Issue | Date

Figure 12.3 Record specification for a printed report

Record Specification

The detailed contents and format of each record must now be specified. For ease of reference, this is shown on a separate form called a record specification (fig. 12.3) cross-referenced to each record identified on the Computer Document Specification. In addition to describing the data items contained in a particular record, this form also allows the format of the record (line or block) to be described. The relationship of data items to each other are described by the level conventions (chapter 10).

An exception to this rule is that the first line of this form is assigned level 01 and refers to the whole line or block of print. This is followed by a description and the position of each data item contained within that line of print, identified at levels 02, 03, etc, as appropriate.

Where a line of print (level 01) is fixed relative to the top of the output document, its line number position is taken from the layout grid counting the first line of the grid as line one, and the appropriate number entered in the 'from position'.

If the position of a record is not fixed in this way, lines are numbered relative to their position in the record counting the first line available for printing within the record as line one. In this case, the number to be entered in the 'from position' may be prefixed with the letter 'R' to indicate a relative number.

Where the position of a line in a record is relative to the last occurrence of a preceding line of variable occurrence, a relative line number prefixed by a 'plus sign' may be entered in the 'from position' to indicate the number of lines from the last occurrence of the variable line.

To indicate the end of a line print which occurs a number of times the final line number is entered in the 'to' position. Where lines occur a fixed number of times, the line number of the last occurrence should be shown, or the letter 'V' where there are a variable number of occurrences.

Lines which occur at regular intervals on a page may be described by a comment immediately below the entry, prefixed by an '*' in the reference number column, eg, '* every fifth line'.

Each record and each data item has been identified by a unique name, and this name is entered in the appropriate space.

The position of each data item within the line is indicated by entering the character position of the first and last character of the data item taken from its position on the layout chart.

Display chart

Figure 12.4 shows a chart which can be used for specifying the visual layout of messages passed between the computer and terminal equipment in the

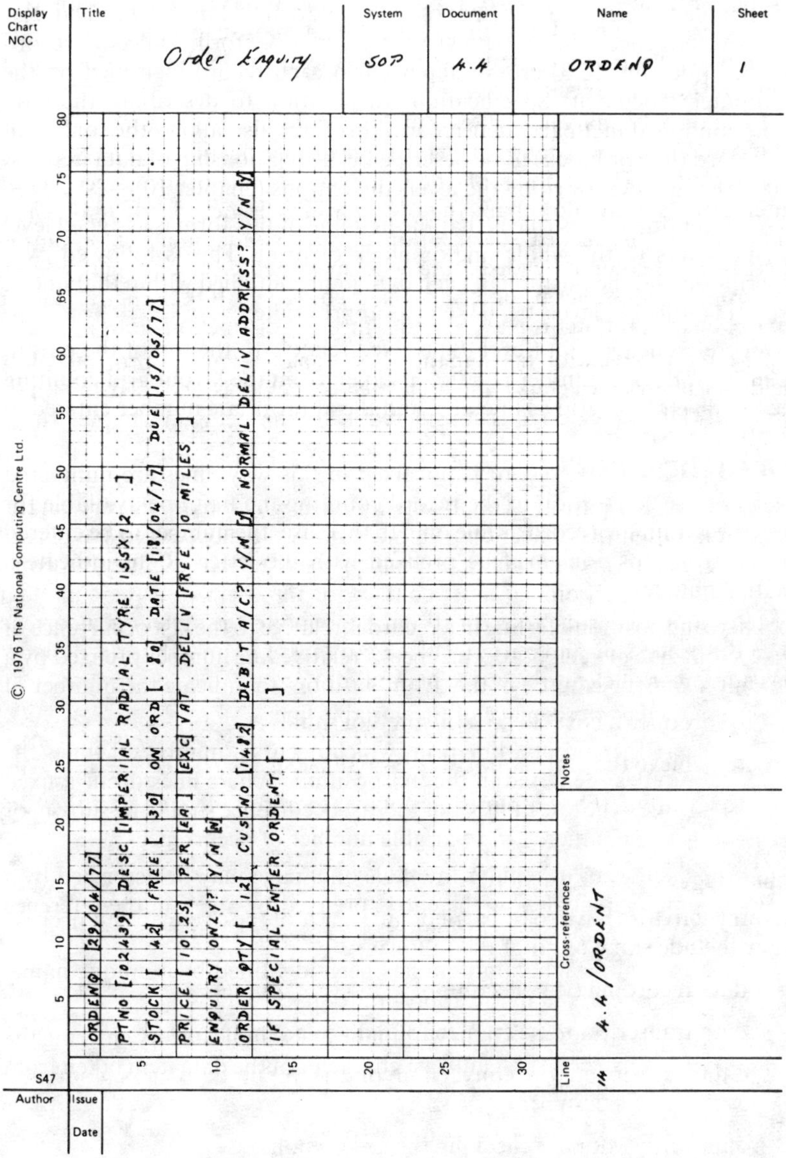

Figure 12.4 Display chart

form of hard copy or screen display. This chart is used to depict the format of the complete message to be input or output. It has capacity for up to 80 characters in 32 lines. If the physical device has more capacity than this, the Print Layout chart can be used; if it has less, then a line can be drawn around the available area.

Each display chart illustrates a message containing a record or number of records, which will be specified on a record specification (fig. 12.5) for the programmer. The linking of these messages may be described in an interactive system flowchart (fig. 13.4) which is described in the next chapter. As this record specification may be used for input or output messages the column headed 'data type' may be used to distinguish input from output messages by entering 'IN' (I) or 'OUT' (O). Similarly, on the layout chart input items may be underlined for distinction.

The main convention in using the display chart is to enter protected data fields as plain characters; and unprotected fields as enclosed in square brackets. Dialogue design is considered in greater detail in Chapter 16.

INPUT DESIGN

Input design is a part of overall system design which requires very careful attention. Often the collection of input data is the most expensive part of the system, in terms of both the equipment used and the number of people involved; it is the point of most contact for the users with the computer system; and it is prone to error. If data going into the system is incorrect, then the processing and output will magnify these errors. Thus the designer has a number of clear objectives in input design:

- to produce a cost effective method of input;
- to achieve the highest possible level of accuracy;
- to ensure that the input is acceptable to and understood by the user staff.

Input stages

Several activities have to be carried out as part of the overall input process. They include some or all of the following:

- data recording (ie collection of data at its source);
- data transcription (ie transfer of data to an input form);
- data conversion (ie conversion of the input data to a computer acceptable medium);
- data verification (ie checking the conversion);
- data control (ie checking the accuracy and controlling the flow of the data to the computer);

Record Specification
NCC

Record description	System	Document	Name	Sheet
Order Enquiry	SOP	4.7	ORDENQ	1

Medium: VDU | Record length: Fixed ☑ Variable ☐ | Record format: Fixed ☑ Variable ☐ | Record size: Words ☐ Characters ☐ Bytes ☐ | File specification refs. | Lay-out chart ref.: 4.4/ORDENQ

Ref.	Position From	Position To	Level	Name In system design	Name In program	Data Type	Size	Alignment	Picture	Occurrence	Value Range
1	2		01	Heading Line			17				
2	2	7	02	'ORDENQ'		0	6		A(6)		
3	10	17	02	Current Date		0	8		X(8)		Valid date
4	4		01	Part No Line			49				
5	2	5	02	'PTNO'		0	4		A(4)		
6	7	12	02	Part No.		I	6		9(6)		
7	15	18	02	'DESC'		0	4		A(4)		
8	20	49	02	Description		0	30		X(30)		
9	6		01	Stock Line			64				
10	2	6	02	'STOCK'		0	5		A(5)		
11	8	11	02	Total Stock		0	4		ZZZ9		
12	14	17	02	'FREE'		0	4		A(4)		
13	19	22	02	Free Stock		0	4		ZZZ9		
14	25	30	02	'ON ORD'		0	6		A(6)		
15	32	35	02	Stock outstanding		0	4		ZZZ9		
16	38	41	02	'DATE'		0	4		A(4)		

S 44
Author | Issue | Date

Figure 12.5 Record specification for a display

- data transmission (ie transmitting, or transporting, the data to the computer);
- data validation (ie checking the input data by program when it enters the computer system);
- data correction (ie correcting the errors that are found at any of the earlier stages).

Not all of these stages need to be present; for example, with on-line data entry, data conversion and verification are not usually necessary. Indeed, one of the aims of the systems analyst must be to select data capture methods and devices which reduce the number of stages so as to reduce both the chances of errors and the costs. Nor need the stages be carried out in the sequence listed; data control and data correction will be involved at several points, and data transportation may occur before data conversion.

Input types

One of the early activities of input design is to determine the nature of the input data. This will have been done partially in logical system design but it now needs to be made more explicit. Inputs can be categorised as:

- external, which are the prime inputs for the system, eg sales orders, purchase invoices;
- internal, which are user communications with the system, eg file amendments, adjustments;
- operational, which are the computer department's communication with the system, eg job control parameters, file names;
- computerised, which are inputs in computer media coming from other internal systems or external systems, eg bank records passed over on magnetic tape;
- interactive, which are inputs entered during a dialogue with the computer.

In addition to identifying the input types the analyst needs to consider their impact on the system as a whole and on other systems. Often inputs to one system can be used as inputs to another; and inputs can be pre-processed or completely raw when entered into the system (this means that clerical procedures must be designed alongside inputs).

Input media

Once the input types and their contents have been examined the analyst can start to think about input devices, of which there is a very wide range. The first classification of input devices might be source-document conversion devices, like card punches, paper tape punches, key-to-tape, key-to-disk, key-to-cassette, and key-to-diskette. These are used for conversion of data from source documents into computer acceptable media.

There are also by-product data-capture devices such as billing machines, accounting machines, and cash registers; these are used to capture data in a computer-acceptable form as a by-product of some other essential operation. And there are direct data capture devices, such as Optical Mark Readers, Optical Character Readers, Magnetic Ink Character Readers, and Kimball and Datatag Tag readers; these devices are linked to the computer and receive the source document directly without any conversion or verification processes.

Another classification is on-line data entry devices, such as teletypewriters, visual display units, data collection devices, audio response terminals, light pens and optical wands, which collect data directly from the source document into the computer one transaction at a time.

Much careful thought has to be given to the choice of input media and devices. Consideration can be given to:

- type of input;
- flexibility of format;
- speed;
- accuracy;
- verification methods;
- rejection rates;
- ease of correction;
- off-line facilities;
- need for specialised documentation;
- storage and handling requirements;
- automatic features;
- hard copy requirements;
- security;
- ease of use;
- environment of data capture;
- portability;
- compatibility with other systems;

- cost;
- etc.

These areas are explained in depth in books on data capture.

Error avoidance

Every effort must be made to ensure that input data remains accurate from the stage at which it is recorded and documented to the stage at which it is accepted by the computer. This can only be achieved by careful control each time the data is handled.

The conditions under which the tasks involved are carried out can affect the legibility and accuracy of the data. For example, dirty and damp conditions in office or factory premises can affect the fitness of both people and machines and consequently the effectiveness of the results; poor form design can lead to a misunderstanding of the instructions or insufficient space on which to write clearly; lack of control can lead to documents being lost or mislaid without the loss being realised. Any measures which are taken to improve these conditions are likely to reduce the amount of errors and loss of data.

The effectiveness of checking data by verification or sight-checking can only be assessed by keeping individual records of the preparations of input data and 'tracing' errors which are subsequently found by the computer, or even later in the system, back to their source.

Error detection

While every effort is made to avoid errors during the preparation of input data, past performance shows that a proportion of errors is always likely to be present. Experience also teaches that the further into the system that errors are found, the more complex may be their effect and the more difficult to correct.

Controls

Control checks can be carried out in various ways, usually at three levels: batch, record and item.

A batch can be any collection of transactions, usually of a regular size suitable for checking and control convenience, or perhaps a total quantity or value processed per period (eg day, week, month). The transactions in each batch are totalled by the computer as part of the input, and then compared with controls previously prepared and fed into the computer with the user input data. Any discrepancy is then shown as an error which must be checked, corrected and re-input.

Errors may be in the control batch total itself, individual records/items incorrectly totalled, keyed or overlooked somewhere along the line of input procedures; even a whole batch of transactions may have been mislaid. The control procedure must therefore be designed to detect errors at every level.

The systems analyst must, as part of the input design activity, design the system of controlling the processing of batches of data. This involves determining the nature of the batches (on the basis usually of location, timing or volumes); calculating the data volumes in relation to time and deciding on the make up of a batch; designing batch control slips and batch progress records; setting up procedures for batching and totalling and setting up procedures for error location and correction. The flow of a batch is illustrated (fig. 12.6).

The batches will be fed into the computer accompanied by a control record, which may contain the following data:

- date of preparation of run or other data relevant to the process and unique to the run (eg, today's date);
- run identification code;
- batch identification;
- number of records or groups of records in the batch;
- control totals eg totals of all values or quantities in the batch or hash totals of product codes or customer codes.

The position of the batch control record, whether in front or behind the relevant batch will also need to be specified.

Variations to this procedure are designed appropriate to the methods of data collection and input processing, eg on-line interactive data entry from visual display, audio response terminals, turnround prerecorded documents.

Data validation

Computer input procedures must also be designed to detect errors in the data at a lower level of detail which is beyond the capability of the control procedures. These are combined with the design of the input process itself, reading the data on the input medium from the appropriate peripheral unit, editing and transferring data to a magnetic medium for subsequent main processing, and sometimes including a partial sorting process.

The validation procedures must be designed to check each record, data item or field against certain criteria specified by the systems analyst for the programmer. Each type of record has codes to be checked for acceptability. As the record type indicates that a certain process or series of processes are to be performed, an incorrect or non-existent code must cause the whole of such a record to be rejected showing the reason by means of narrative or a set of error codes designed for this purpose.

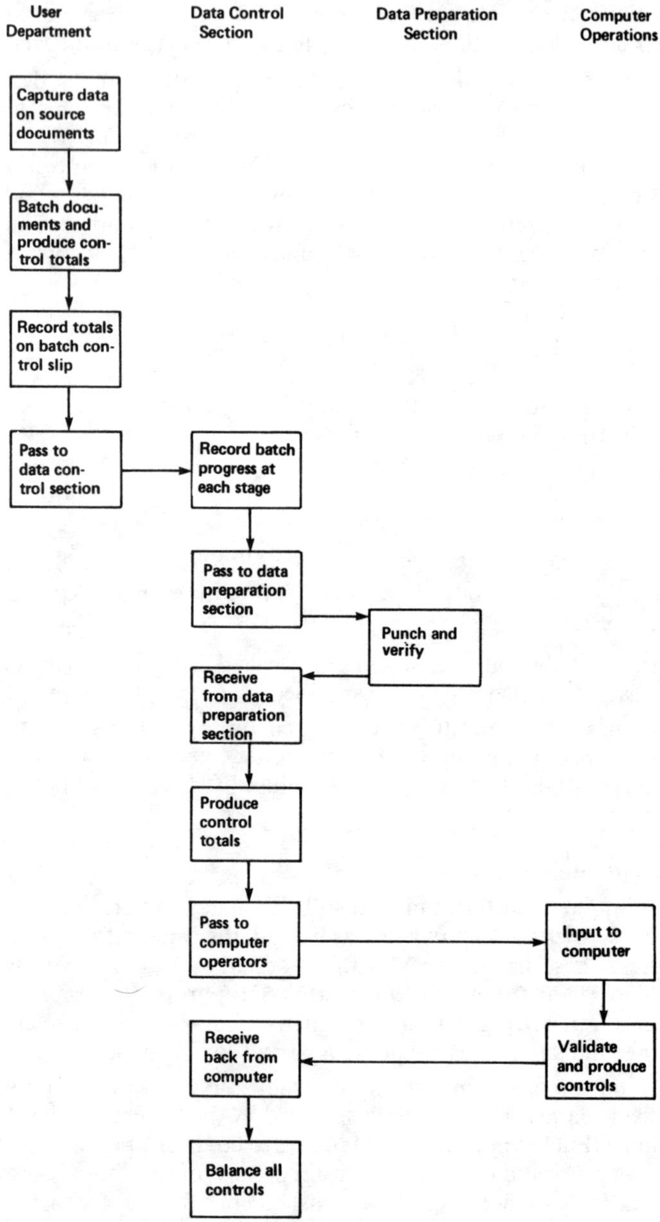

Figure 12.6 Flow of batch

Each item can also be checked against its unique value, eg composition, size, format. The COBOL 'picture' is normally used for this item representation as characters, words, bytes, packed or unpacked. Data items are given specific names, comprising letters, numbers, symbols; they can be constants (item names which remain unchanged) or literals (which specify precisely the value or content of a constant). These items can be checked as equal to, less than, or greater than specific values; they can be checked against tables or formulae.

The contents of some fields are dependent on the contents of others, ie they must be checked in combination with other fields as well as individually (eg if field 1 is A, then field 2 must be in a range 2000–3000).

It is not possible to discuss every type of check which may be used, but the type of data item itself normally suggests the type of check required. The data specification documents shown in this book provide columns in which to specify check values (eg value range, picture, alignment, data type and size of data item on the record specification). As the validation features for many input procedures can be made the same within an organisation, the general validation logic can be prepared as a standard package or routine program for which the designer specifies data requirements as appropriate.

There are of course limits to which these checks for accuracy can be taken. For example, a customer account number 12345 can be checked that it is numeric, equal to 5 decimal characters in length and is in a range 00001 to 99999; but its precise value within that range cannot be checked by any of the methods described above. To overcome this problem, especially for checking key fields, a method is used called check characters or check digits which is explained in Chapter 17 in some detail because of its importance.

Error handling

Once the system has detected an error, careful procedure's must be followed in re-submitting the correction. Ideally errors should be printed out in a form on which they can be corrected and from which they can be punched again. The error reports should be clearly identified and distributed quickly to the recipients so that re-input can take place swiftly if necessary.

Data acceptance

Data which passes the tests for validation is deemed to be acceptable and is written from main memory to magnetic media for subsequent processing.

In one type of system this accepted input data may be processed without waiting for the rejected data to be corrected and re-input. Another type of system may be such that the converse of this applies. The correct action to be taken will depend on the effect that not processing the rejected data will have upon the system compared with, say, the delay of waiting for the re-input of corrected data.

How much data should be rejected? Just the item, field or character apparently in error, the whole record, or the batch? In the interests of the extra time that would be involved if whole batches or records were to be re-punched and re-input, alternative methods such as writing the suspected batches/records to a suspense file on magnetic media and only re-inputting the actual corrected data, then clearing the data on the suspense file to the accepted data file, may be preferable.

In real-time processing three attempts will normally be made to correct an error by attempting to re-input the message correctly, after which, the message will be accepted, or rejected completely. The effect of a real-time rejection will depend on the nature of the message. An aborted enquiry may merely inconvenience the enquirer while a transaction which would have modified the condition of a master record is more serious, eg a stock item could then be shown to have an incorrect balance or appear to be out of stock, when it was not (or vice-versa), to a following transaction or enquiry; or in another context, a customer's credit limit may appear to have been (incorrectly) exceeded.

This situation can also apply to batch processing, but the time interval between runs is not so critical and may allow time for the position to be corrected before any serious ill-effects can arise; they are more easily controlled.

Once data has been validated, it may of course be changed, or edited. Fields which have been used for error control may perhaps be discarded; constants can be inserted; fields can be expanded and changed; other editing features may include:

- stripping off redundant data;
- radix conversion;
- storage mode conversion;
- field extension, truncation or amalgamation;
- justification or alignment;
- switch or flag setting;
- repetition of fields;
- preliminary arithmetic operations.

In other words, input data should be designed to suit the user until it gets into the computer; then it can be changed to suit the computer.

Other input design considerations

Some of the other considerations which the analyst must take into account are:

- the nature of input processing, eg complex, involving all inputs, or separate for each input;
- flexibility and thoroughness of validation rules;
- handling of priorities within the input procedures;
- use of composite input documents to reduce the number of different ones;
- relationship with other systems, eg can the validation procedures carry out checks that might be required by another system which makes use of the input data?
- scheduling of input runs in case of large rejection rates at validation;
- forms design, to ensure accuracy and efficiency of input;
- relationship with files, eg types of data which can be stored or which need to be input at each run.

INPUT SPECIFICATION

Input files can exist in document form before being input to the computer, in which case their specification is described on a Clerical Document Specification (fig. 7.8, vol. 1).

When input data is converted into computer files, (card, paper tape or magnetic) they are described by a Computer File Specification (fig. 10.4) and their detailed contents by a Record Specification (10.6). These data specification forms apply to all computer data, not only to input data. As an example of the record specification being used to specify an input file, a card input file is shown (fig. 12.7).

SUMMARY

Output and input data will have been identified in outline in earlier stages of the system design activity; this chapter is concerned with their detailed physical design. The approach to output design is very dependent on the type of output and the nature of the data, but special attention has to be paid to editing and totalling of output data. Once the output medium has been chosen, detailed specification of output documents can be carried out. Input design is rather more complex since it involves procedures for capturing data as well as inputting it to the computer, and these will vary

Record Specification NCC

Record description	System	Document	Name	Sheet
Order Batch Control	SOP	4.5	OBATCC	1

Medium: Punched card — Record length: Fixed ☑ Variable ☐ — Record format: Fixed ☑ Variable ☐ — Record size: 80 — Words ☐ Characters ☑ Bytes ☐ — File specification refs.: 4.2 /ORDINP — Lay-out chart ref.:

Ref.	Position From	Position To	Level	Name In system design	Name In program	Data Type	Size	Alignment	Picture	Occurrence	Value Range
1	1		02	CDCODE		C	1		X		10
2	17	19	02	NRORDS		C	3		999		001-999
※				No of orders in the batch							
3	20	25	02	QTYORDS		C	6		9(6)		000001-999999
※				Total quantity ordered in batch							
4	26	31	02	DESPNO		C	6		9(6)		100000-999999
※				Serial no. of Control Sheet (see 4.1/DESPSHT)							

S 44 Author — Issue — Date

Figure 12.7 Record specification for a punched card file

depending on the type of input. Careful design of input stages, after the input medium has been chosen, involves attention to error handling, controls, batching and validation procedures. The inputs then need to be specified in detail both as clerical documents and as computer input files. Output and input design, therefore, involves considerable attention to detail and integration with file and procedure design.

13 Computer Procedure Design

INTRODUCTION

A computer procedure is a series of operations designed to manipulate data to produce output from a computer system; the procedure may be a single program or a series of programs. Computer procedure design is treated separately in this book for ease of understanding, but would be carried out in an integrated way with the other aspects of physical design of the computer subsystem.

The detailed design of computer procedures follows acceptance by management of an outline design proposal. The aim now is to design procedures at lower levels of detail which will define the detailed steps to be taken to produce the specified computer output (or intermediate stages) from the initial input of data. When complete, these procedure definitions together with data specifications are organised into a specification for programmers from which the required programs can be written. The extent to which the specification of individual programs is done by the systems analyst depends on the standards of each installation; in some, the systems analyst is responsible for detailed specification of each program; in others, the programmers design individual programs from an overall procedure specification. Even in the latter case the systems analyst will have a major influence over programs because of the responsibility for file design and identification of system procedures.

Some operations must be performed in a certain way because they are inherent in the system itself; others are performed according to the processing method adopted, eg, sequential processing with a serial device, or random processing with a direct access device.

DESIGN TOOLS

Various tools are used by systems analysts to specify computer procedures, eg narrative, flowcharts and decision tables. These have already been described (Chapter 7) and will be discussed here only from the viewpoint of their specific use in computer procedure design.

Flowcharts

Flowcharts of computer procedures can be drawn at various levels. In theory there is no limit to the number of levels to which a system can be divided; the convention in this book is for the overall system to be defined by a System Outline (fig. 7.7 vol. 1) for an overview of the total system, and by a System Flowchart (fig. 7.4 vol. 1) for a more detailed examination of both computer and clerical procedures. System Outlines and System Flowcharts can be drawn for the complete system or for subsystems; for example, there may be a System Flowchart for the complete production control system, and a series of System Flowcharts, one for each subsystem, such as Order Entry, Stock Control, Production Scheduling, at a lower level of detail.

Once the overall system or subsystem flowchart has been produced, the boxes which represent computer procedures have to be defined in detail and a number of types of flowchart are used for this.

Computer Run Chart

A Computer Run Chart is a way of showing the inputs, processes and outputs in the computer part of the system in a logical structure. It can be regarded as a master plan of the computer subsystem, but cannot be used to show details except by cross reference. Figure 13.1 shows a Computer Run Chart for a computer despatch procedure. It shows the inputs, master files, processes, transfer files and outputs, in that order, from left to right across the sheet with the flow of data shown by the direction of lines and arrows. Inputs, outputs, and master files symbols bear the reference of their appropriate data specifications, and the process symbols of their procedure specifications at a lower level. The symbols used are part of the set shown in figure 7.3 (vol. 1).

In a sequence of processes, execution of the first is followed by execution of the subsequent ones in the order indicated. Such a processing sequence executed at predetermined periods, eg daily, weekly, monthly, implies batch processing; if executed as an immediate response to a single transaction it implies real-time processing. In batch processing, the same procedures (eg, input validation) may be included in a daily run chart, as well as a weekly (input – validate – update – print) run chart.

A Computer Run Chart is used to group logically-related sequences into specific procedures; it is not necessarily expressed in terms of programs; and a process symbol on a Computer Run Chart does not necessarily refer to a program. This run chart will subsequently be replaced by a programmer-produced run chart which will specify programs. These may

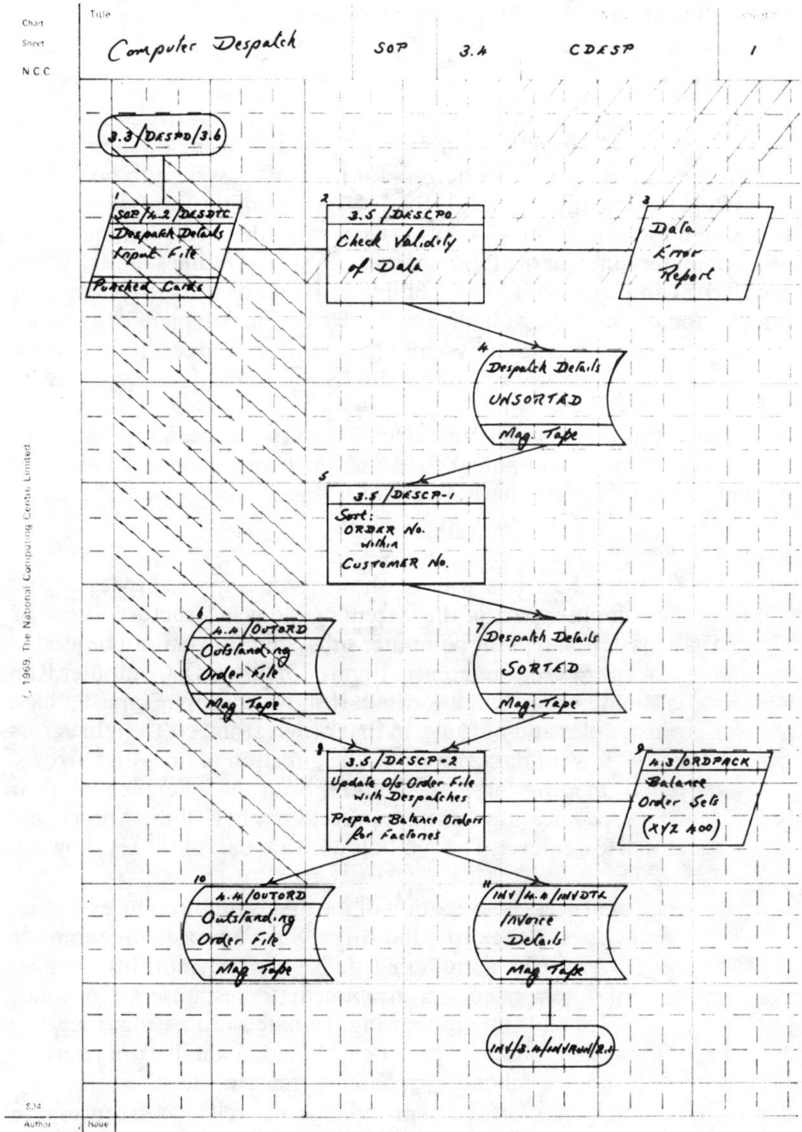

Figure 13.1 Computer run chart

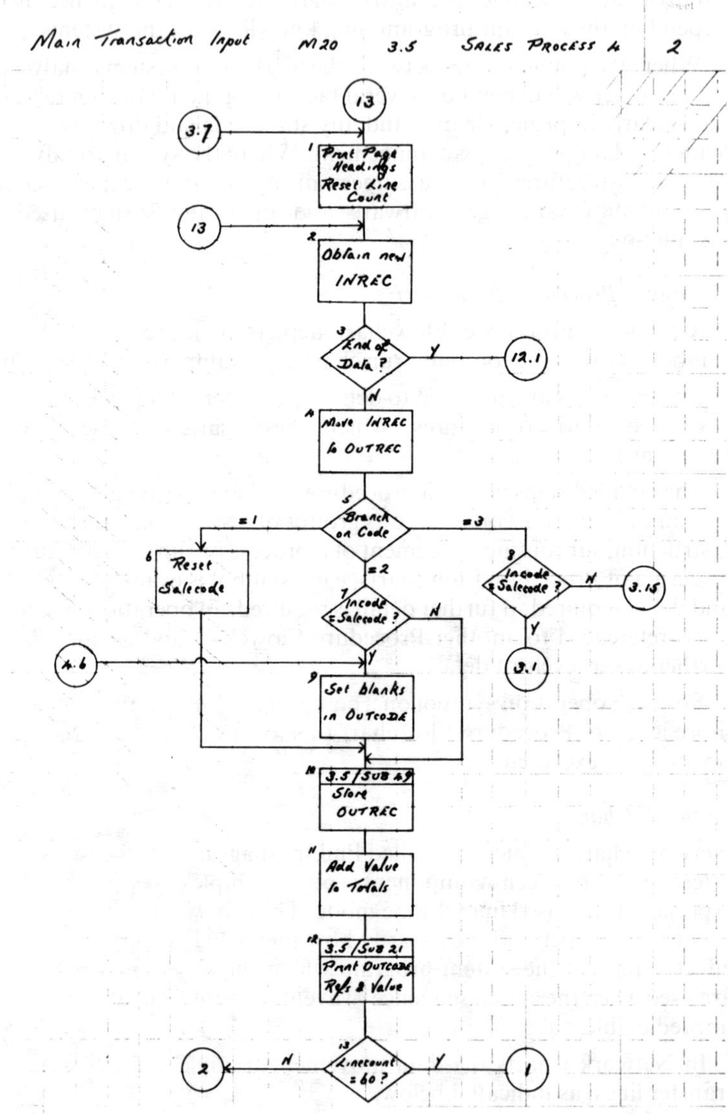

Figure 13.2 Computer procedure flowchart

coincide with the procedures as specified by the systems analyst, or may be subdivided or aggregated, into programs. Any coincidence may well depend on the relevant programming knowledge of the systems analyst.

Where the boundary lies between the work of the systems analyst and the programmer will depend on several factors, eg their working relationship, particularly in project teams, and any standards laid down by the project leader or data processing management. Where the systems analyst designs program structures, this assumes an appropriate experience of the programming language, software and operating system used in the installation.

Computer Procedure Flowchart

A Computer Procedure Flowchart depicts in more detail the process symbols shown on a computer run chart. An example is given in figure 13.2.

The processes are analysed to define their internal logical steps and then organised into procedures which best satisfy these operational requirements.

The detailed steps within a procedure level may be as small as individual machine instructions or as large as groups of instructions (forming a macro instruction, subroutine or segment of a process or run). To define the levels of design of Procedure Flowcharts, each symbol is examined for relevance and detail required; if further detail is required, an operation symbol can be cross-referenced to another Procedure Flowchart and expanded into still further lower levels of detail.

For each operation symbol on a computer run chart, there will be one or more levels of Procedure Flowchart to show in detail how that computer operation is executed.

Network Charts

Network charts (sometimes called Phillips diagrams) are used to depict the interactions between components of a complex system or program, especially those used in real-time mode. They show, at a general level, how events *can* happen, not necessarily how they *will* happen. They provide in effect, a map of the system/program which shows all possible routes, and are used when the actual sequence of events cannot be readily specified or is unpredictable.

In Network Charts the data movement triangle is used to qualify the transfer lines, as indicated below.

Direct Control – General ————————→

– transfer of control from one process to another;

Direct Control – Temporary

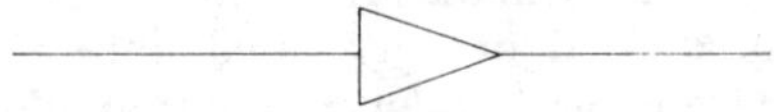

– transfer of control from one process to another, with direct return to the calling process when the called process is completed;

Direct Control – Permanent

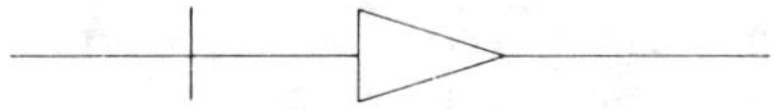

– transfer of control from one process to another with no guarantee of return to the calling process;

Indirect Control

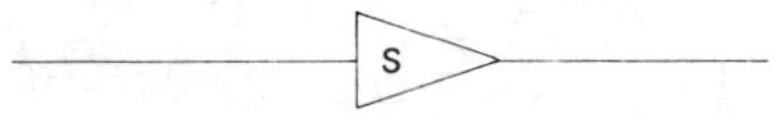

– activation of one process by another via an operating system or scheduler. An alphanumeric symbol within the triangle indicates the type of activation, eg 'S' for 'Scheduler';

Interrupt

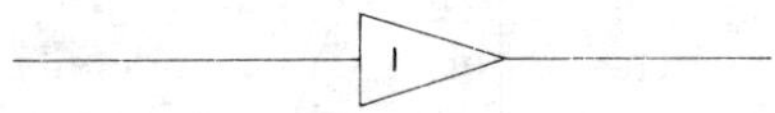

– activation of a process by means of an interrupt generated externally, eg by a peripheral device.

It can be seen (fig. 13.3) that arrows are used more frequently in network charts than in flowcharts, and that two lines are used to indicate two-way flow. Also it should be mentioned that these charts must always be supported by lower-level documentation which specifies data and processes involved.

Interactive System Flowchart

An Interactive System Flowchart is used to depict the sequence of events in an on-line system where the user is in control of the program at execution time and can enter data between execution steps or determine the sequence of operations. The Chart (fig. 13.4) is divided into three main areas showing, from left to right, terminal user procedures, terminal formats and computer procedures. Each area may be subdivided into columns to show different levels or locations of operation. Procedure boxes are cross-referenced to their detailed procedure specification; terminal formats are cross-referenced to display charts (fig. 12.4).

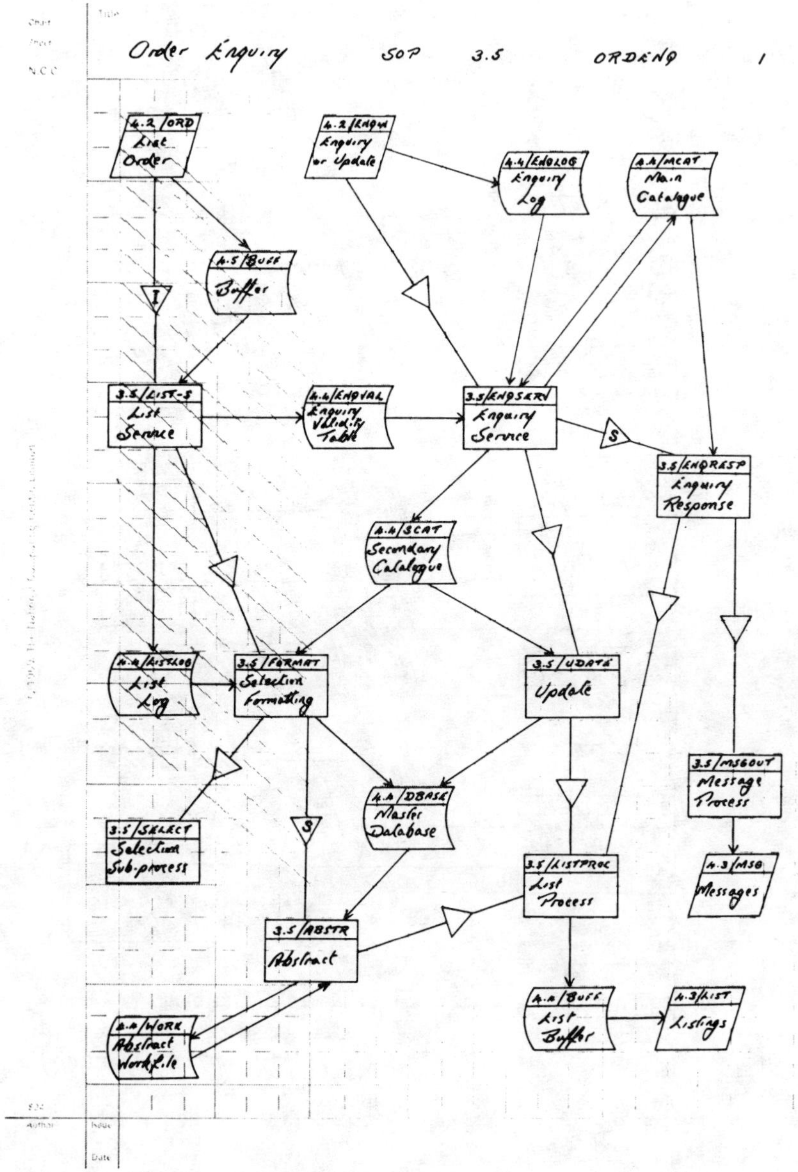

Figure 13.3 Network chart

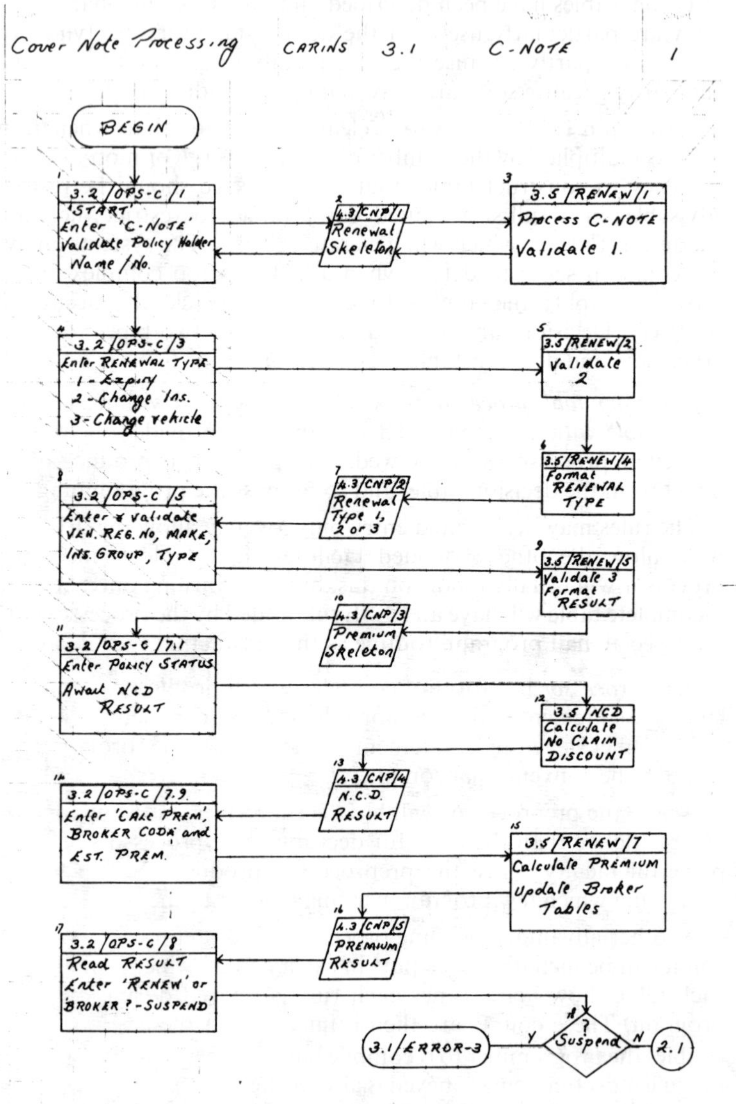

Figure 13.4 Interactive system flowchart

Decision Tables

Decision Tables have been described and illustrated in Chapter 7 (vol. 1). They are particularly useful at the design stage for specifying computer procedures, partly because they are readily understood by programmers and partly because software is available for handling them.

A Decision Table is likely to be clearer than a flowchart when the number of rules multiplied by the number of conditions is 6 or more. However, the tendency to construct tables that are too large should be resisted. It is advisable, in a full-size limited entry table, to restrict the number of conditions to four, which will generate 16 rules. The systems analyst must use common sense to detect when a Table is too complex and when it should be split; the aim is to communicate clearly, accurately, and concisely. Decision-table software usually is of two types: decision-table processors and decision-table preprocessors.

A *decision table processor* is usually incorporated within a language or is compatible with a compiler. This means that, provided the rules for that particular processor are followed, it is possible to produce object code direct from the decision table, and so by-pass the programming activity.

The rules may say 'limited entry only' or may allow the 'ELSE' rule, or may allow the other extended facilities. If a table is incomplete, the processor will usually report on this, but as a warning only. Sometimes an incomplete table will have an ELSE rule added by the processor, with a call to a special 'halt program' routine as the only action.

A *preprocessor* is a form of compiler which produces a series of source language statements from decision tables. These can then be added to other programmer-generated statements, and the whole program compiled through the conventional compiler.

Again, the preprocessor will have certain rules, but sometimes they are less restrictive than the rules for decision table processors. This is a very powerful facility, since the preprocessor produces perfect source code, something which few programmers manage first time.

Another advantage of some processors and preprocessors is that a rule count can be included. This puts in extra coding which simply counts, for each table, how many times each rule is obeyed during one run of the program. These counts are then printed out at the end of the run. This enables the programmer to recompile but with the rules resequenced so that the rule most frequently obeyed is always the first ('Else' cannot be first, but must always be last) and so on in descending frequency of use.

Thus, a tested program can be given a large set of live data to handle, the rule count taken, and the program tuned to give a maximum run-time efficiency. On the recompilation, the rule counters and logic are removed, because they slow the program down. This can be repeated from time to time (say yearly, at most) to ensure that efficiency is maintained, and that the pattern of data has not altered significantly.

TYPES OF COMPUTER PROCEDURE

Computer processes can basically be classified into six types:

- validation;
- sorting;
- processing files on serial access devices;
- processing files on direct access devices;
- processing a database;
- printing.

The processes of validation, sorting, processing and printing will be included in most systems. For example, the computer run chart (fig. 13.5) shows a typical batch processing system.

Validation

Data enters the computer-based system via validation procedures (Chapter 12). Often they are catered for by a generalised input validation package tailored to the needs of the particular system. The major decisions at the validation stage are concerned with handling errors and distribution of data.

Error handling

Error procedure must be specified in detail showing decisions, actions and exceptions. There are various ways of handling errors open to the designer, including:

- rejecting the item of input and processing the next item;
- writing an error record onto an error file;
- signalling the operator by a message on the typewriter or other device;
- not updating or processing the file or batch in which the error occurred;
- dumping the contents of the file where the error was detected;
- going back to an earlier stage in processing and starting again from that point;

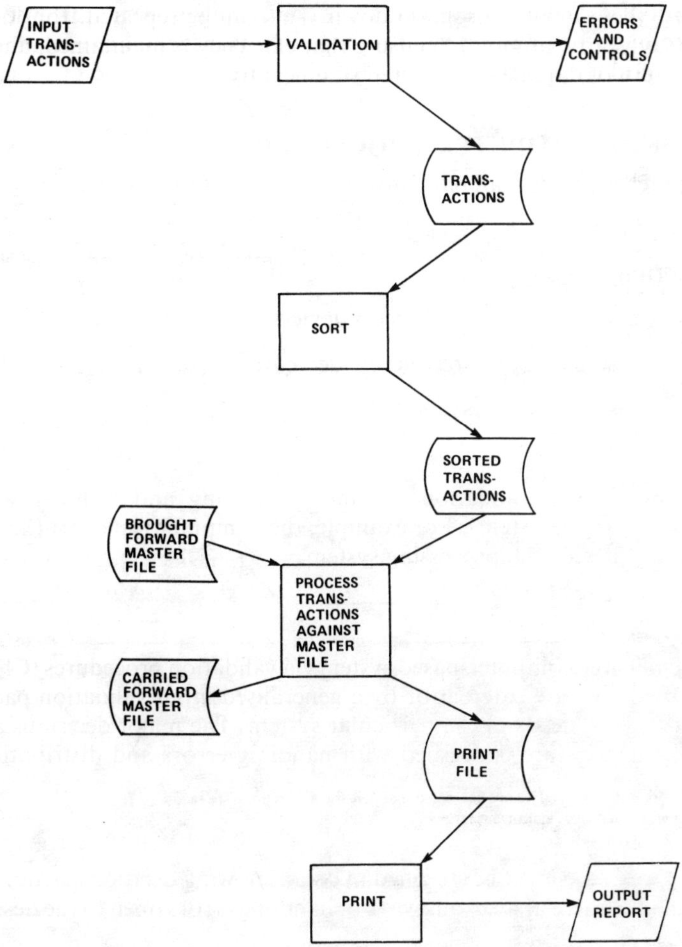

Figure 13.5 Computer run chart for typical batch processing system

- halting the program or run;
- allowing the system to correct the error internally;
- sending signals to other programs affected by the error;
- manual procedures for correcting the error.

If the input data must be correct before continuing processing, the designer has the option of writing valid records to a transaction file on magnetic media and bringing that file forward to the next run of the program (fig. 13.6).

First alternative

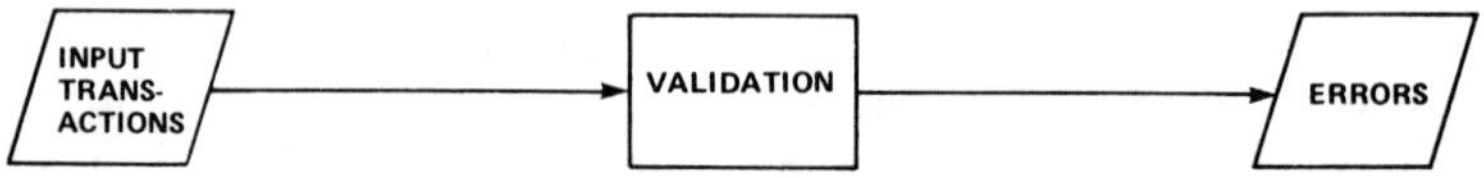

The errors in the input data are corrected and the full set of input transactions are read again.

Second alternative

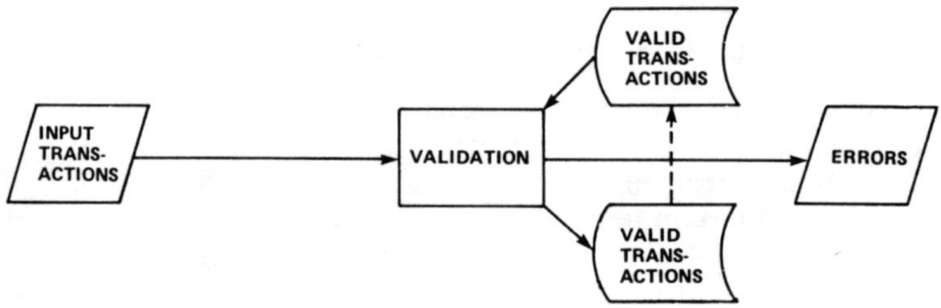

The errors in the input data are corrected and just the corrections are read again to be merged with the valid transactions.

Figure 13.6 Alternative approaches to validation

Data distribution

Once validation is complete, data has to be distributed to the relevant files. This can be a complex task: in some systems, communication of input data to files may require several runs or complete passes of the whole system before the full effect of the input has been achieved. Furthermore, a transaction may trigger feedback into the files. In a real-time system this will all happen at once, but in a batch-processing system, it may take days.

For example, in a batch processing stock control and accounting system, if a customer's order were to be passed through the stock file updating routine before the customer file updating routine, it might be found that the customer's credit limit was exceeded and that no orders should be processed until outstanding balances were cleared. This would mean that the alterations to the stock records would have to be reversed on the next run of the whole system. Conversely, if the customer file processing routine came first, including invoicing, an invoice could be prepared and then, on reaching the stock file updating routine, it might be found that there was no stock. There would then have to be feedback to credit the customer account on the next run of the system.

In addition, the format of transaction records must be such that they are processed in a logical order (eg it is necessary to process transactions which maintain master files before updating transactions and queries). In a batch processing system, the code structure of the different types of input will usually take care of this. In a real-time system where transactions are handled individually, ways must still be built into the system to ensure that transactions which are entered independently will not be logically incompatible, cancel each other out or be processed in reverse order.

Thus considerable thought must be given to the output transactions from the validation process and their sequencing.

Sorting

Data frequently has to be sorted before a further phase of processing can be carried out. The time spent sorting, which can represent 40% of the total processing time, is really non-productive and must be reduced as much as possible. It is important, therefore, that the systems analyst understands the implications of the various methods of sorting, and the facilities provided in sorting software.

Sorting in the general sense means arranging data according to a particular rule or pattern. In this context, sorting consists of arranging logical records in sequence according to the key in the record.

Merging may be defined as the production of a single sequence of records dependent upon some rule of order from two or more sequences already in order to the same rule.

In data processing, what is commonly described as sorting, is really a combination of sorting and merging. It may take place off-line, where the information to be sorted is contained on punched cards or computer readable documents.

The bulk of sorting however is carried out on-line and may be classified as:

- internal sorting;
- magnetic tape sorting;
- direct access sorting.

All computer sorting takes place in main storage, but the term 'internal sorting' is usually reserved for that which makes no use of peripherals. When the whole of the data to be sorted cannot be held in core, some of it must be held on backing storage, either magnetic tape or a direct access device. In both of these cases a proportion of the data may be read into main store, sorted and the various portions or strings of sorted data are then merged, and either held in main store ready for the next operation or written to magnetic backing storage.

Sort keys

Any record which is to be sorted must contain a sort key. This will indicate the desired position of the record within the file arrangement. It may be a special field within the record provided for this purpose or may more commonly be an integral part of the record such as an identifier (eg customer code, product code).

All sorting processes depend upon the way in which data is stored by hardware and the instructions available to the machine. Keys can be expressed in binary or as characters, both of which are treated as strings of bits.

If character strings are specified, the sequence into which records are sorted will depend upon the internal machine code used to hold the characters. All machine codes are in an alphabetic sequence for upper case alphabetic characters. If lower case is also specified, this will also be in alphabetic sequence, but not in sequence with the upper case letters. Numeric characters and special symbols may appear at any point in the code and if these are to form parts of sort keys, the systems analyst must fully understand the internal coding, in order to specify the sort correctly.

To ensure that records are sorted correctly, numeric key fields must be zero filled on the left and alphabetic characters must be left justified. In other words, key fields must be of fixed length. Furthermore, they must be in a fixed position relative to the start of the record and variable field must be accommodated following key fields.

Sorting can usually be carried out in either ascending or descending sequence and multiple keys are usually catered for in sort programs. For example, it is possible to sort addresses alphabetically to the sequence of street within town, within county. Character string keys may be of any reasonable length, but both long keys and a large number of keys will increase the time required for sorting.

Methods of sorting

Internal sorting is carried out by standard sort programs or modules which generally offer little or no choice in the methods of internal sorting employed.

Magnetic tape sorting normally involves reading a small group of records from an input file, arranging them in a particular sequence, referred to as a 'string', within the core store and writing the sequenced records away to a work tape. The process is continued until all the records have been sequenced and written away in short strings. Then two or more short strings are read into core store and merged into a longer string. These are written away and the process continues, generating longer strings, until the whole file is in sequence.

Magnetic tape sorting then can be regarded as being carried out in two phases, pre-string and merge; but if the amount of data to be sorted is greater than can be held on one reel, each reel is sorted separately by pre-stringing and collating, and then all the reels are merged. Collating and/or merging may require a number of passes of the magnetic tapes before the final sorted file is generated. Usually, the more work tapes that are available, the fewer the passes and the quicker the sort.

Direct access sorting is required when it is quicker to sort a direct access file and process serially rather than process randomly, or when reorganising a file. The techniques used are similar to those used for magnetic tape, consisting of identical pre-stringing and similar collating. The basic difference is in the collate phase. It is possible to split a disk file into many subfiles, each of which can be used in the same way as a magnetic tape. Thus, the method of merging is not limited by hardware availability. In practice, the number of subfiles is limited to about 15, since read/write head movement time between the various subfiles becomes too great beyond that number.

A totally different technique known as 'tag sorting', is sometimes used with direct access media. With this method, the key is extracted from each record and, together with the address of the records, these tags are formed into a subsidiary file which is then sorted, often in main store. This sorted subsidiary file thus provides an index to the main file, which for many purposes, need not be sorted any further. If, however, it is essential to have the main file in sorted order, it is a comparatively quick process to rearrange it with the aid of the sorted tag file. The comparative efficiency of record and tag sorts will depend upon the amount of main store available and the ratio between record size and key size.

Sort software

All manufacturers provide sorting routines of various types. The way in which these operate and their relative efficiency vary a great deal. Internal sort subroutines are usually written in assembler language and can be called by a user program which will supply the necessary parameters. Sort and merge programs are free-standing programs which take either a magnetic tape or direct access file as input, sort on the appropriate media and output a completely sorted file. Direct access sorts will usually accept input and output on tape, but the reverse is not true. These programs are controlled by parameters, the data of which is read from cards or paper tape.

Sort/merge generators contain similar routines to the sort programs but allow the user to add own coding to them, usually in assembler language. Facilities available are:

- first pass own coding;
- last pass own coding;
- equal record own coding;
- sentinel and label own coding.

At first pass own coding, each record is presented to the user for processing before it is passed to the sort. It can be used, for example, to insert leading zeros into numerical keys. At last pass each record is presented to the user before it is written to the final output file. It could similarly be used to remove the leading zeros inserted on the first pass. Alternatively, instead of writing the record out to a file, it could be output to the line printer. In this way, a separate print program might be saved.

Most standard sort routines do not make provision for sequencing records which have identical keys. These records will normally be output in random order, within key sequence. If the user wishes to do anything special with them, equal record own coding provides this facility. For example, in sorting a customer transaction file, the user might wish two or more orders from the same customer to be treated on a value basis or may wish to combine them into one record.

Label records in a file which is being sorted are usually ignored by software. Where these records hold control information, the user will not wish to lose this and must therefore, arrange to process it using the sentinel and label own coding facility.

Processing files on serial access devices

The only way to process a file on a serial medium (eg magnetic tape) is by each record or block of records being read into main store one by one in the order in which they occur on the medium. Access to a specific record is only possible by first reading all the records preceding it. Since it is not possible to overwrite the exact position of a previously read record, the records on the file can be changed only by copying them serially from one device to another and at the same time incorporating any changes to the relevant records on the new file. This applies, for example, to a file of input data where faulty data has been rejected and a new file of validated data produced, or where a 'sort' programme produces a new file of records in a predetermined sequence. It applies particularly to master files on magnetic tape, when the records which they contain need to be changed and brought up-to-date.

File maintenance and updating

Processing transactions against master files is known as *file maintenance* with reference files, and *file updating* with dynamic files. There are usually far fewer changes to be made to reference files than to dynamic files, which are changing all the time as a result of updating transactions. Also changes to reference files often require more complex processes than do changes to dynamic files.

When files are being maintained or updated it must be possible to identify on both transaction and master files:

- each individual record on the file, usually via the key field;
- each different type of record within the file, (eg a parts record, an outstanding order record, a purchase order record);

Transaction records must also be classified by type into:

- insertions, ie new records which need to be inserted in sequence in a master file (eg new employees, new customers);
- deletions, made when records cease to be required, (eg employees leaving employment, suppliers no longer trading). They are removed by not copying them to the new file; this prevents unnecessary growth and redundancy;
- amendments, required from time-to-time to change the data values in some fields on a master file to reflect the correct situation (eg, customer's change of address or correction to an incorrect balance);
- updating transactions, ie actual business events, such as the values of products despatched, products returned, money received, discount allowed, carriage paid, in a transaction to update a sales ledger master file; the quantity and value of outstanding parts orders, parts received, parts issued, parts returned, to update a stock master file.

Notification that a change is to be made to a master file arises when an amendment or transaction record (which for simplicity we will refer to as a transaction) is processed by the computer. The only way to change a master record is to read the transaction, then to find the master record and change it accordingly.

If the records were held in random order on magnetic tape, this would necessitate reading the first transaction, searching along the master file tape until the keys of both records matched, making the appropriate change to

Title	System	Document	Name	Sheet
SERIAL UPDATE	ORDENT	3.4	SERUP	1

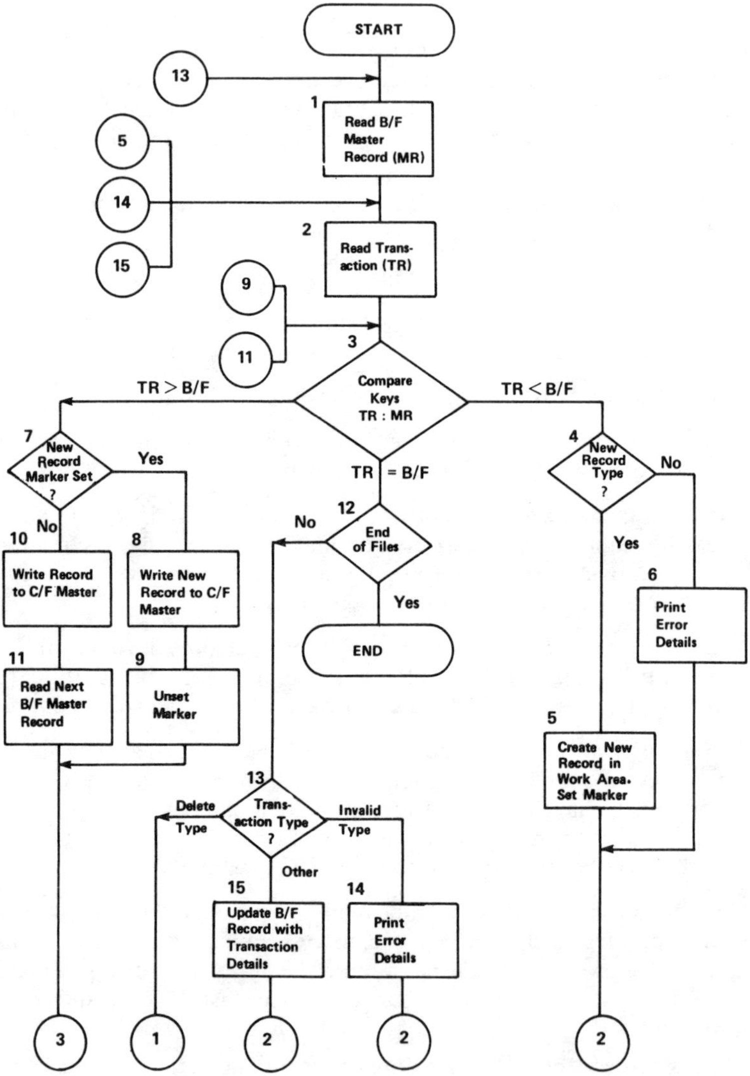

Figure 13.7 A simplified serial sequential updating procedure

the master record and then rewinding the master file tape to start the search from the beginning for a match to the second transaction key and so on. This would clearly be time-wasting and nonsensical. The obvious solution is to sort the master file records into key sequence when it is first created, and then to sort the transaction file records into the same sequence before they are matched against the master file.

The processes of maintaining and updating magnetic tape files are the same. Although there are normally many more updating transactions than there are maintenance transactions, both functions require the matching of transactions against master records. Reference and dynamic master files may be kept as separate files and maintained and updated separately, or they may be combined into one master file containing both reference and dynamic data in one record, maintained and updated in the same program run. (Even with combined records updating and maintenance may be separate.)

When maintaining and updating a master file in one run, reference data amendments and updating transactions must be sorted to the correct transaction type sequence within record sequence as follows:

- deletions;
- insertions;
- amendments;
- updating transactions.

The process of changing the records of master files works on the principle that a new updated master file of one cycle (sometimes referred to as the carried forward (C/F) file), will become the input master file (brought forward (B/F) file) of the succeeding cycle. In the following example (for which a similar procedure flowchart is shown in figure 13.7, both transaction and B/F files are organised sequentially and terminate with a dummy high key value record (say 9999) to simplify the logic of the end-of-file procedure. Using single reel files, the following procedure requires three magnetic tape decks to read two input files (ie one transaction file and one B/F master file), and to write one output C/F master file; and assumes four areas of memory, one for each file and one work area. This is not the only logical solution, but it is typical.

The following is a narrative version of the flowchart (*compare both to test the facility of the flowchart*):

1 read the next B/F master record into main store memory;
2 read the next transaction record into memory;
3 compare the keys of both B/F master and transaction records in memory;

4 if the key of the transaction record is less than the key of the B/F master record, the transaction must either be the insertion of a new record which is not on the master file, or the transaction type is incorrectly coded. Test the transaction type code.

5 if the transaction type code is valid (ie the transaction is the insertion of a new record), assemble the details of the new record in a predetermined work area and 'set' a marker to indicate this. Then go to (2) and read the next transaction;

6 if the transaction type is an invalid code, print the details as an error. Then go to (2) and read the next transaction;

7 if the transaction record is greater than the B/F master record, test if a new record marker has been 'set' by procedure (5);

8 if the answer is yes, write that new record from the work area in memory to the C/F master file;

9 'unset' the new record marker, and go to (3);

10 if the test at (7) indicates that the new record marker has not been 'set' by (5) above, write the record in the B/F input area to the C/F master file. This may be a record from the B/F master file for which there is no transaction, in which case it is copied without change, or the B/F record may have changed with one or more transaction details. In this example, a record key 9999 would signify that the last record on the transaction file had been read, which will in effect force the program to read and copy the remaining records on the B/F master file to the C/F master file until the key 9999 is read on the B/F master file;

11 read the next record on the B/F master file and go to (3);

12 if the keys of both records at (3) are equal, test for end of files condition (keys equal 9999). If so, the program and file control software will then close files and terminate the run;

13 if the 'equal key' records are not end of file records, test the transaction type code. If the transaction code is a delete type, this indicates that the record identified on the transaction must be deleted from the master file. Go to (1), read the next record from both the B/F master file and the transaction file. Reading these into memory, in fact, overwrites the record to be deleted which cannot then be written to the C/F master file;

14 if the transaction type code is not accepted as valid, the whole transaction must be rejected and the details printed as an error. Go to (2) and read the next record on the transaction file. (The B/F master record will be subsequently copied unchanged to C/F master file);

15 otherwise, the transaction must be an amending or updating transaction and the contents of the B/F master record in memory changed or updated accordingly. There is often more than one transaction for the same master record. Go to (2) and read the next record on the transaction file. Repeat this cycle until the keys of transaction and B/F master records are not equal.

The above procedure is repeated until all the records from the input files have been read and processed.

By combining file maintenance and updating into one computer run the latest amendments to the reference data take place in step with the updating of the latest events. Other advantages of the combined approach are:

- only one matching pass of the main file is required;
- sequence of processing is controlled;
- the operation of the computer system is easier.

The advantages of the separate approach are:

- the design to incorporate maintenance of reference file sometimes involves complex procedures;
- less coding is required in each computer program and so sometimes more comprehensive procedures can be accommodated;
- it may be possible to use a standard program for maintenance of reference files;
- easier programming.

Whatever the balance of the particular arguments in a specific situation, it is advisable to study very carefully any proposal based on the use of separate file maintenance runs.

Buffering

Designers of both computer hardware and business systems are continually seeking ways to exploit the potential of the central processor, particularly to reduce the disparity between the processing speeds of the central processor and the peripheral units. The use of separate peripheral control units and multiplexer channels which allow data to be transferred simultaneously along different channels and the development of multiprogramming, virtual memory and spooling, are examples of this continuing trend. One of the techniques which makes use of such developments is known as 'buffering'. In the example in figure 13.7 of a serial sequential update, a record is read, processed in memory and then written out in its updated form to the C/F master file; then the next record is read; and so on.

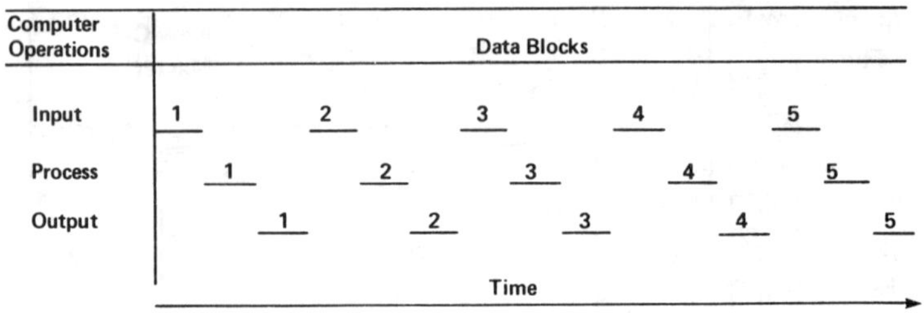

Figure 13.8 Non-simultaneous operations

Figure 13.8 illustrates the inefficiency of the serial nature of these operations, ie that the central processor is not being used during the input and output operations, and the total time of this cycle therefore is the total of the three operations in series. To avoid this serial operation, predetermined areas of memory, called buffers, are allocated, each of sufficient size to receive the input or output blocks of data. The result is illustrated (fig. 13.9). There is more than one method of buffering, using one or more buffer areas, but they have the same aim – simultaneity of computer operations.

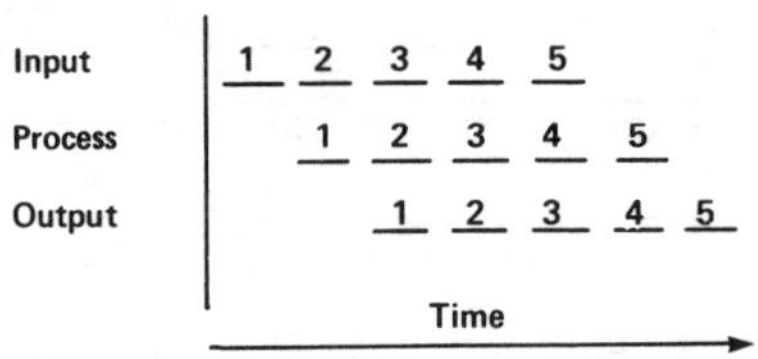

Figure 13.9 Simultaneous operations

Double buffering uses two buffer areas for input and two for output, into or out of which blocks of data alternate. The first block of data is read into the first buffer and processing of this data takes place simultaneously with the reading of the next block of data into the second buffer. Once processing of the first block of data has been completed, the second block can be processed immediately. As the first area is now free, the third block of data is read into this area during processing of the second block. Thus the two buffers are used alternately for accepting input.

A similar method is used for producing the output file. As records are available for output, they are moved first into one area then into another for writing to the new master file. Figure 13.10 illustrates double buffering used for updating a master file with a transaction file and shows the use of work areas.

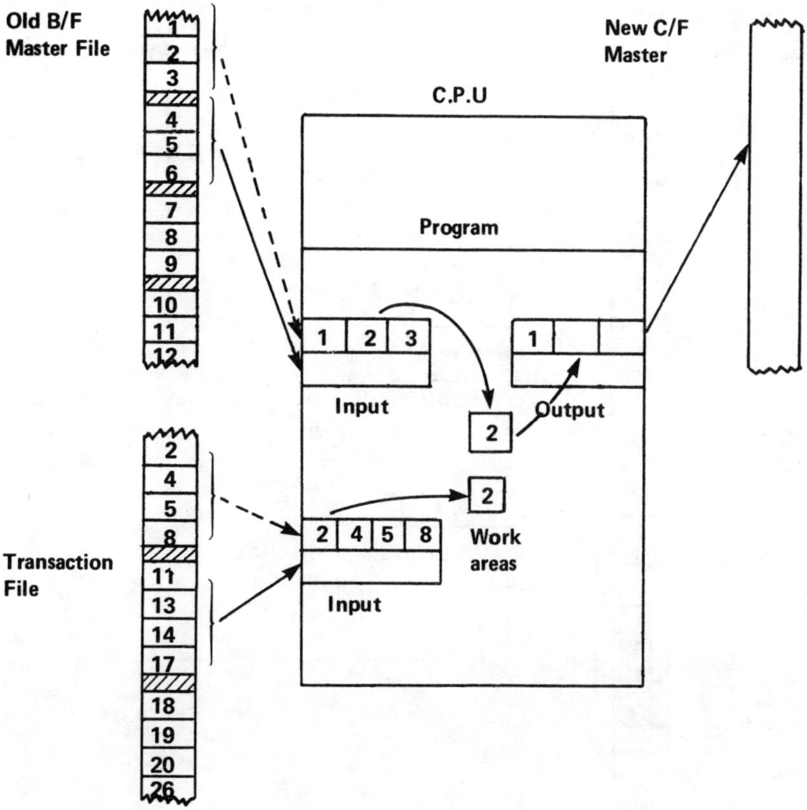

Figure 13.10 Double buffering

Records may be grouped into blocks on the file. File handling software will cater for the packing and unpacking of blocks of records, handling them to the program one at a time as they are requested and accepting them from the program one at a time, as processing is complete.

Where single records, rather than blocks of records, are being handled, the system of using two buffers for input and two buffers for output can be combined for economy of store into one treble buffering system (*see* fig. 13.11). Here, each of the three buffer areas is used in rotation for input, processing and output. While a record is being read into buffer area A, the

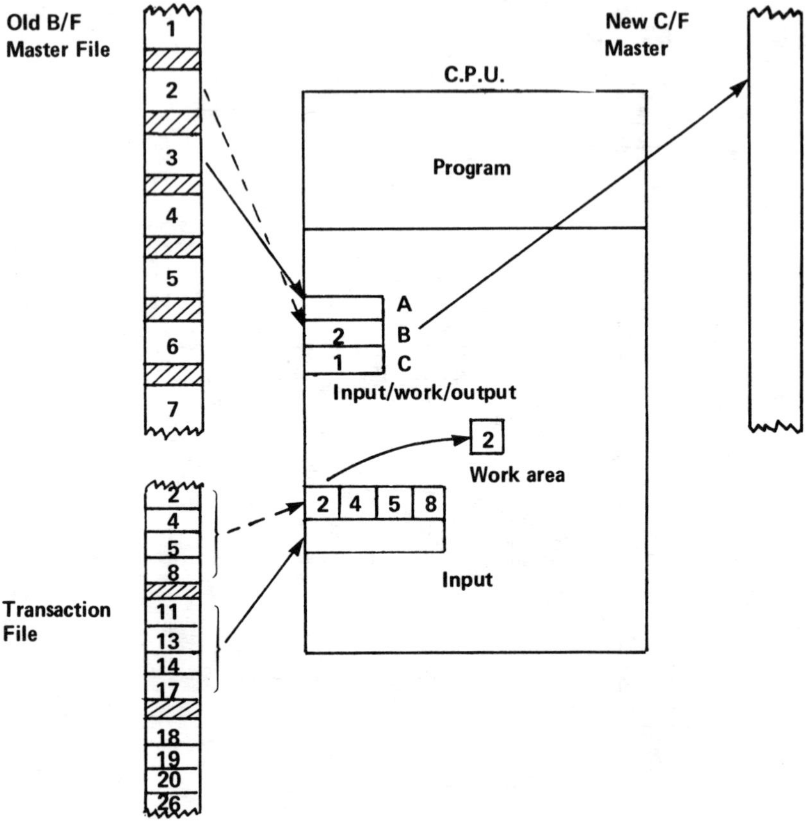

Figure 13.11 Treble buffering

record in buffer area B is being processed and the record in buffer area C is being written to the output device. Once processing of the record in buffer area B is complete, this area then becomes the output area. Buffer area C is available as the next input area, and buffer area A contains the next record for processing.

Buffer areas must be provided for each file to be processed. The amount of memory available for this purpose may well govern the size of blocks that are used on the file. A compromise has to be reached between maximising block size for efficiency of transfer and minimising block size for efficient use of main memory.

Processing large and complex files

Where the size of a file is such that it can be updated simply in one process, a major problem is to determine how many functions can be accommodated in the procedures which are to handle the file. If, however, the file is very large or complex, or its processing is too complex to be accommodated into one process, it may be necessary to adopt more involved file structures. There are various methods of processing in this situation, some of which are described below:

Abstraction: if some parts of the data on a file have to be accessed more frequently than its whole, it may be worthwhile to generate an abstract or summary file. For example, to produce a distribution network plan using a customer file, an abstract file could be prepared consisting only of customer number and each customer's delivery codes, omitting the bulky details of names and addresses, price codes, etc. This technique would also allow the abstracted data to be in a sequence which differed from that of the main file, if required.

Changes file: if the distribution of activity over the file population is skew, eg if 80 % of the changes affect only 20 % of the records on a master file, only those records which are actually updated are written to output. Subsequent runs then require, in addition to file updating transactions, the B/F master file and the changes file, producing a progressively-updated changes file containing all those records which have moved.

Dead file: in some situations records may die or become redundant when they reach the position that no further activity can be expected; for example, on a local authority's 'rate collections' file, once an account is in balance it will not move until the next rate account is billed. In this situation, it may be advantageous to carry forward only those accounts which are still alive, keeping a dead file for processing at some subsequent date.

Extraction: if a large file is subject to a very complex procedure, it may be advantageous to separate the processes of matching and updating. In this way, the matching program, which may be very simple, can be made to handle large blocks of data on the main B/F and C/F files, so increasing the effective file passing speed. Those records which are to be updated are written onto an intermediate file with a smaller block size. The updating work is then carried out in a separate run. The updated records written by the second program are merged back with the main file on the next run of the extract program.

Handling transfers

A problem which often arises is that of transferring a record or part of a record from one position of the file to another. There are three basic methods of handling transfers:

- delete the original record and re-insert with all the data;
- change the key and then re-sort the files. (This is only suitable for large numbers of transfers in one run);
- delete the record and write it to a transfer file. (This file is then sorted to update the master file on the next cycle).

When records have been transferred from one position to another, it is usually worthwhile to leave a 'shadow record' in the old position. This re-directs any fresh data which may be wrongly referenced during the time following the transfer. Re-directed data will have to be treated in one of the ways suggested above.

A similar problem occurs when two or more files are related. For example, if a statistics file is updated by details of transactions posted to the sales ledger, it becomes necessary to decide what action to take in the event of an amendment to the ledger. To allow such amendment to modify the statistical records would involve considerable programming effort, but if the amendment is not carried through, it will not be possible to balance the statistical and ledger totals at the end of the year.

Processing files on direct access devices

The way in which a file is processed on a direct access device need not bear a direct relationship to its organisation. For instance, one can have:

- sequential processing of sequential files;
- random processing of sequential files;
- sequential processing of random files;
- random processing of random files.

There are five basic methods of processing for files held on direct access devices: serial, sequential, selective sequential, binary search, random.

Serial processing

This implies the processing of records in the sequence of their physical location. In other words, each access after the first is the next highest address within the file, regardless of the content of the records held there. The method is similar to the processing of a file held on magnetic tape. If it were to be used on a structured file, no account would be taken of indexes or keys and overflow conditions would be ignored.

Serial files are always processed serially, since there is no other method of locating the data within the file. If specific records were required, all records would need to be searched. This is generally the method of processing adopted for transaction files.

Sequential processing

In this method, records are processed in logical key sequence. Processing can start from the beginning of the file or from a specified record within it and can stop at any point. The organisation of records on a file for sequential processing need not, of course, be serial, provided that there is some means of addressing, using either indexes or a direct addressing algorithm.

This method of processing would be used for sequentially organised files where the hit rate is high and one or more records per page is required for processing. The type of procedures required would be similar to those described earlier for magnetic tape files.

Selective sequential processing

This implies the processing of records within a file, still in specific sequence, but only processing the records that are required. Inactive records are passed over, and the next one required is processed without handling each inactive one. The method may prove to be better than sequential processing of the whole file (on the one hand), and direct processing of selected records (on the other), but it will, of course, entail sorting the input file to a specific sequence. It is normally used where, on average, less than one record per page is required during the run, and is the most commonly used method in practice on commercial master files.

To use selective sequential processing, there must be some means of addressing the records required, either by use of an index or by calculation of the address.

Binary search

The binary search technique starts by reading the middle page of a sequential file, and testing to see whether the record keys in that page are equal to, less than, or greater than the key required. If one of the retrieved keys is equal to the required key, the search is over; if the retrieved keys are higher than the required key, the process is repeated for the first half of the file; if the retrieved key is lower than the required key, then the process is repeated for the second half of the file. (The technique can also be used for searching indexes.)

The formula for required page number is:

$$\text{Page no.} = \text{low page no.} + \frac{(\text{high page no.} - \text{low page no.})}{2}$$

Example: To read a file with page nos 001–999, the first page read is:

$$001 + \frac{(999 - 001)}{2} = 500$$

if the keys found are lower than the required key, the next page to be read will be:

$$501 + \frac{(999 - 501)}{2} = 750 \quad \text{and so on.}$$

Random processing

Random processing enables direct retrieval of particular records by presenting the appropriate keys; no sequence need be maintained. The method may be used on any structured file whether sequential or random, where transactions are processed in the sequence in which they are received and each unit of stored data has an equal chance of being retrieved for the next transaction. The method is normally slow when compared with, say, selective sequential processing, due to large head movements. In addition, when used for non-sequential access to a sequential file by use of index tables, the index will need to be consulted for each access. Large indexes are required and index search time is high. On the other hand, access is rapid and volatile files are easily handled. To minimise head movement, it may be necessary to store high activity (frequently occurring) records adjacent to each other, using a measure of activity as part of the record key, when the file is created.

In random processing, master records after updating are always written back to the same location from which they were read. Transactions do not need to be sorted prior to processing, as access to a particular master record is direct. Since transactions can be dealt with as they occur, it is particularly suitable for real-time operations.

The flowchart in figure 13.12 describes the procedure to update a master file on a direct access device with a randomly organised transaction file (which may be a batched input file or a real-time message). The transaction file contains a dummy high value key to signify the end of the file. The master file is organised as a random file, with a key transformation algorithm (KTA) used to convert the key of each record to a direct access address. When a synonym is found, a search is made for the nearest vacant address in which to start a new record. A narrative description of the flowchart follows (again, compare for both methods, the ease of recording the procedure as a designer, and then its comprehension by a programmer).

(1) Read a transaction, (2) test for end of file, (13) if end of file, close files and terminate processing.

(3) If not end of file, transform the key into a direct access address using the KTA.

(4) Test if the resultant address is within the range of the file.

(14) If it is outside the range, carry out the error routine, go to (1) and read next transaction.

(5) If it is within the range, read the master record at the resultant address and compare the keys of the master and transaction records.

(6) If the keys are equal and (15) the transaction code does not indicate the creation of a new master, then (16) update the master record according to the transaction detail. Go to (1) and read the next transaction. (15) If the transaction code does indicate the creation of a new master record when one already exists, then carry out the error routine (14), go to (1) and read the next transaction.

(6) If the keys are not equal, and (7) a synonym pointer is present, (8) extract the address from the synonym pointer and read the master at that address (5). If no synonym pointer is present and (9) the transaction code is not to create a new master record, then (14) carry out the error routine, go to (1) and read the next transaction. (9) If the transaction code does indicate the creation of a new master record, (10) check if all vacant locations have been filled with records – check if 'FF' (File Full) switch is on – if it is, go to (23) on sheet 2, carry out file full routine and go to (1) on sheet 1 and read next transaction.

(10) If 'FF' switch is not set, then (11) check if this location is vacant; if it is vacant, (12) write new record in this location, go to (1) and read next transaction; if it is not vacant, (2/1) then a search must be made for the nearest vacant location, starting with the current address position. Search alternately forwards and backwards from the current position. In this example, this is done by using a counter with two switches, one to indicate that the upper limit of the file has been exceeded and the other that the lower limit has been exceeded.

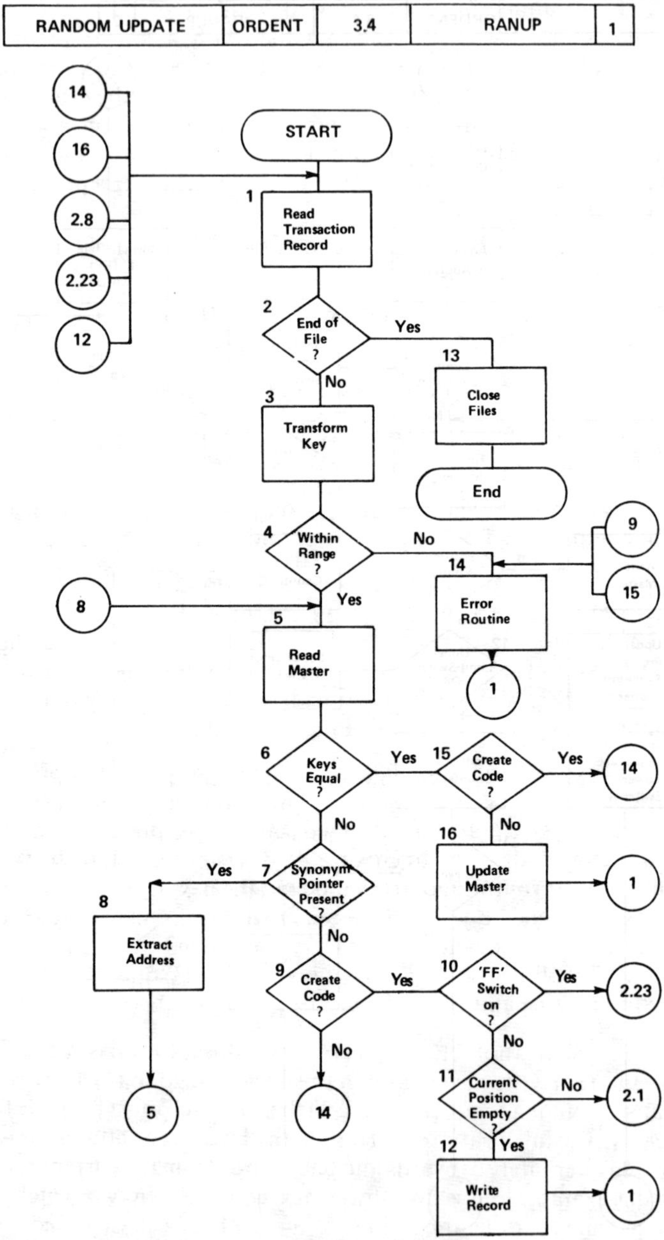

Figure 13.12 A random updating procedure

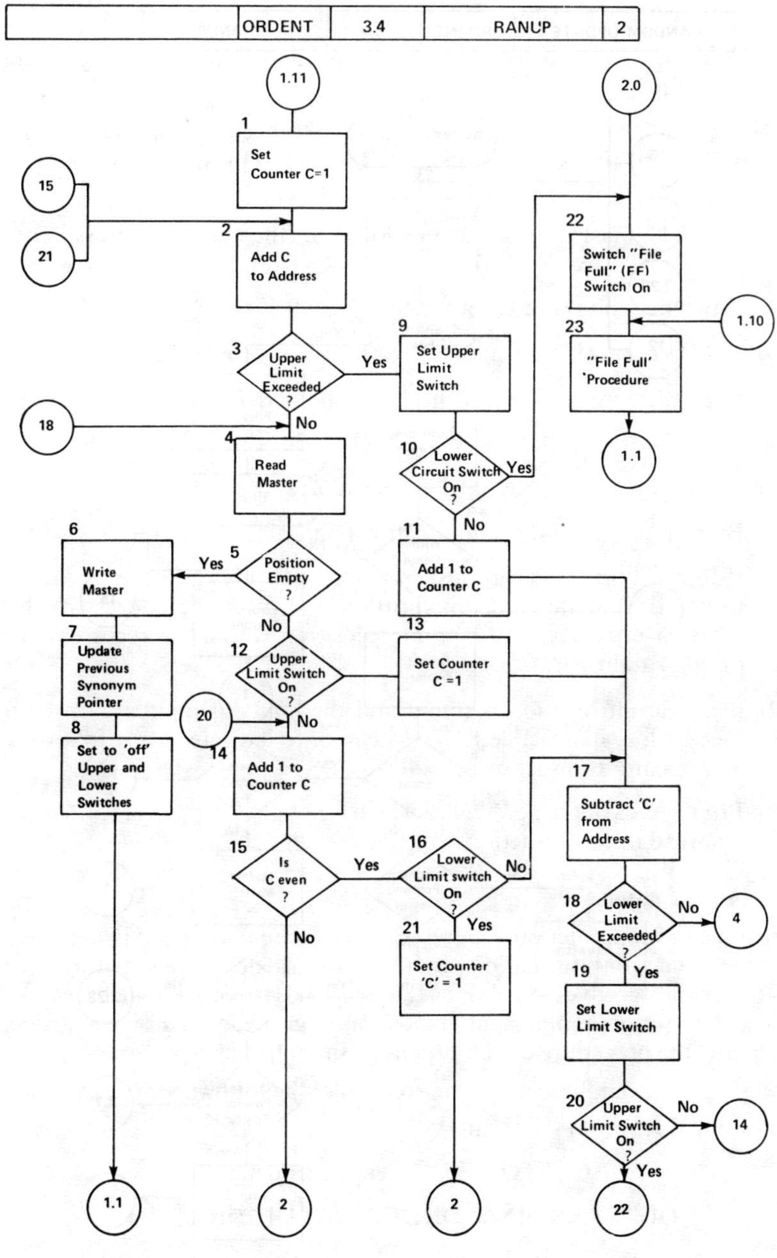
ORDENT
3.4
RANUP
2
1.11
2.0
1
Set Counter C=1
15
21
2
Add C to Address
22
Switch "File Full" (FF) Switch On
1.10
23
"File Full" Procedure
1.1
3
Upper Limit Exceeded ?
Yes
No
9
Set Upper Limit Switch
18
4
Read Master
10
Lower Circuit Switch On ?
Yes
No
6
Write Master
5
Position Empty ?
11
Add 1 to Counter C
7
Update Previous Synonym Pointer
12
Upper Limit Switch On ?
13
Set Counter C =1
8
Set to 'off' Upper and Lower Switches
20
14
Add 1 to Counter C
17
Subtract 'C' from Address
15
Is C even ?
16
Lower Limit switch On ?
21
Set Counter 'C' = 1
18
Lower Limit Exceeded ?
4
19
Set Lower Limit Switch
20
Upper Limit Switch On ?
14
1.1
2
2
22

(1) On sheet 2, set a value of '1' in a counter 'C'.

(2) Add the contents of counter C to the address of the current location............

The remainder of this procedure is summarised below to assist the reader who is following the procedure in figure 13.12, step-by-step, through all the various paths.

(4) Read the master record at this location (the current address + contents of 'C', say, 1000 + 1 = 1001).

(5) If the location is not vacant then (14) add '1' to 'C' (1 + 1 = 2).

(15) If 'C' is even, (17) subtract 'C' from address (1001 − 2 = 999).

(4) Read the master record at this location (999).

(5) If this location is not vacant, then (14) add '1' to 'C' (2 + 1 = 3).

(15) If 'C' is not even, (2) add 'C' to the address (3 + 999 = 1002).

(4) Read master record at this location (1002).

(5) If this location is vacant, (6) create the new master record from the transaction, write it in this location and place the address of this location in the synonym pointer position within the record at the original address.

(22) If no vacant location is found and the limits of the file have been exceeded, switch a 'File Full' switch on, to indicate to any following 'new record' transactions.

(23) Print details of the new record which is not placed, go to (1) on sheet 1 and read next transaction.

Processing a database

The design of computer procedures in a database environment is simplified by the presence of information provided by the logical data structure. One of the major objectives of a database approach is to simplify the process of application programming, and in most cases, given a well designed logical database, the procedure design process is simplified also.

To illustrate database processing, consider the following problem.

A printed report in the form:

CUSTOMER customer details

OUTSTANDING ORDER VALUE £xxxxxx·xx

is required for each customer in a database of the following format.

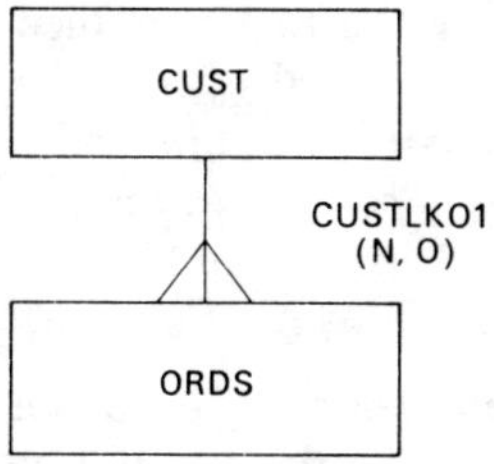

The format of CUST records is:

Start pos'n	Length	Format	Description
0	8	Char.	Control Info.
8	6	Char.	Record Key
14	8	Char.	Next Pointer (to 1st ORDS record
22	36	Char.	Customer Details

The format of ORDS records is:

0	6	Char.	Key of Owner
6	8	Char.	Next Pointer
14	24	Char.	Order Details
38	8	Packed	Order Value

Each customer record is linked to a group of order records using the database technique of chaining. By following a pointer field in a customer record, it is possible to find the first order record on its chain. The next order record on the chain may then be found by following a pointer in the first order record, and so on, until the last order record on the customer's chain points back to the customer record.

The values of each order are to be totalled for each customer. This involves reading each CUST record, and for each record found, the relevant 'chains' of order records must be read, totalling the order value.

The three decision tables in figure 13.13 illustrate how this may be done.

Printing

The function of a 'print program' is to output the results of processing contained on a magnetic tape or disk output file. Usually print programs are provided as part of the computer manufacturer's software; sometimes printing routines are included in a process program. In either case, the printing process will handle records each consisting of one line of print plus control fields which define stationery movement and types of print record. The print records must be presented in the sequence which is required for output. Some of the other points which the analyst must bear in mind with print programs are:

- need for dummy printing at the beginning of the program to allow alignment of stationery, especially if it is preprinted;
- optimisation of printing speeds by careful design of line lengths, lines per page, skipping and spacing;
- accurate estimating of print volumes for timing purposes (and for ordering preprinted stationery);
- printing documents two-up or more to make best use of line space;
- use of output controls like document, page, and line counts;
- output validation before printing to ensure credible data.

One of the major considerations for the systems analyst must be whether to print output data or to produce it on microfilm in view of the increasing cost of paper and the improving facilities for handling microfilm.

DESIGN CONSIDERATIONS

Having examined the tools of computer procedure design and the types of computer procedure, it is now necessary to consider the various factors that are involved in designing these procedures.

Storage constraints

The systems analyst should be aware of those features of the computer which affect the use of main store; on a word-based machine, whether the facilities for dealing with character strings make heavy demands on coding space or processor time; on a byte or character-based machine, whether there is a word substructure. Programming convenience may be such that

	START			Comments
C	'SIGN ON' SUCCESSFUL?	Y	N	Sign on to SCHEMA
A	IGNORE AND FINISH	–	X	Error condition – finish
	REFER TO 'CUST' FILE	X	–	Read the first customer record
	READ FIRST 'CUST' RECORD	X	–	
	GO TO READ1	X	–	

	READ1			Comments
C	END OF FILE?	Y	N	
A	FINISH	X	–	Finish if end of file
	SET UP CUSTOMER DETAILS FOR PRINTING	–	X	
	ZEROISE 'VALUE' ACCUMULATOR	–	X	
	EXTRACT POINTER	–	X	Find and read first order record on chain
	READ 'ORD' RECORD POINTED TO	–	X	
	GO TO READ2	–	X	

	READ2				Comments
C	END OF CHAIN?	Y	N	N	
	SUCCESSFUL READ?	–	Y	N	
A	IGNORE AND FINISH	–	–	X	Error condition – finish
	ADD 'VALUE' FIELD TO ACCUMULATOR	–	X	–	Maintain values total and find next order record on the chain
	EXTRACT NEXT 'ORD' RECORD POINTER	–	X	–	
	READ 'ORD' RECORD POINTED TO	–	X	–	
	GO TO READ2	–	X	–	
	SET UP ACCUMULATOR CONTENTS FOR PRINTING	X	–	–	Print the details of customer and outstanding order value as required after last order on chain
	PRINT CUSTOMER DETAILS AND OUTSTANDING ORDER VALUE	X	–	–	
	REFER TO 'CUST' FILE	X	–	–	
	READ NEXT 'CUST' RECORD	X	–	–	
	GO TO READ1	X	–	–	

Figure 13.13 Three decision tables illustrating database processing

the existence of word and half-word boundaries should be taken account of when making assumptions about record layouts.

Modern operating systems make significant demands on main store and their presence must be allowed for. Where it is possible to delete certain functions from the operating system, thus effecting a saving in storage space, the significance of these options should be appreciated.

The block size chosen on the file medium will affect programming, since main storage must be allocated to whole blocks of data which are in transit between file medium and main store. The block size assumed when making timing estimates should be realistic in terms of the main store available and the total demands to be made on it.

The choice of single buffer or double buffer working has a marked effect on storage, although, in general, there is no advantage to be gained from fully buffered working if the run is processor-dominated. Buffer requirements may be reduced by storing information in a compacted form, but the process of converting back to a format in which the information can be processed may be very time-consuming. Somewhere there is a balance. Where it is possible to nominate one, two or three buffer areas for a file this may be of particular use; for example, where a list requires a significant processing time, and the activity of the file is very low. A large number of buffers can then be employed with advantage to maximise the amount of overlapped file time, by allowing a queue of blocks to build up while the list is being processed.

For efficiency of processing, the system designer must consider the frequency of use of any specific path through a procedure, and where possible this should be indicated to the programmer. Where special procedures are to be accommodated on backing storage, the frequency of occurrence must also be carefully estimated. If, for example, a routine of three hundred instructions occupying 1k bytes of storage is held on a disk with an access time of 100 milliseconds and a transfer rate of 100kbytes per second, a total of 110 milliseconds is required to transfer the routine. While 10 occurrences of that procedure would absorb only 1·1 seconds, 1000 occurrences would require some two minutes, and 10000 – twenty minutes, which may not be acceptable.

Attention needs to be paid to the use of peripherals in each program. Parkinson might have said that 'Programs expand to fill the configuration available'. It is important to remember that although six magnetic file units may be available for use, there will be occasions when only five are operational. It is advisable, therefore, as far as possible to use less than the full configuration available. If more than one peripheral device is available for a specific function, then the various factors which apply to the choice of input, output or storage will need to be taken into account.

Software

In designing a system, the designer should be aware of the capabilities and limitations of software packages. Application and utility packages can do a lot of the work for the designer, speed implementation and cut down programming costs. On the other hand, there may be limitations placed on the design by the operating system in terms of the number of peripherals that can be used, use of main store, etc. In extreme situations, software may require designing if not provided by the manufacturer; occasionally, manufacturers' software may require alteration.

Language

The programming language chosen will have an effect on the amount of main store required. In general, low-level languages give more efficient programs than high-level languages and require less main storage, but take longer to write. It is unlikely however that the analyst will have any choice in the language to be used as this tends to be governed by installation standards, depending on the expertise available, the availability of compilers, and the facilities of the language.

Restart facilities

Restart facilities are generally provided as a function of the operating system. For runs requiring any significant amount of time, it will be desirable to incorporate into the procedures facilities for the run to be restarted in the event of a machine failure. Check points should be provided in the run, at which the contents of the main store are dumped onto magnetic disk or tape. It is then possible to ensure that no more than one interval between check points must be re-processed in order to recover the position in the run at which a failure occurred.

The procedure on restarting is to notify the restart routine of the last check point reached by the procedure in the aborted run. All files are then aligned to the point at which the last dump was made. The dumped store is reloaded and the procedure re-entered at the exit of the check point routine. Check point intervals may be based upon:

- end of reel;
- end of output type;
- time intervals;
- record counts.

A consideration governing their frequency, as so many other things, is cost. To be balanced against the cost of re-processing are the costs of:

- additional programming for recovery procedures;
- increased processing time for recovery procedures with consequent reduction in total throughput;
- off-line storage and media for dumped files.

Sorting

The major aim must always be to avoid sorts as far as possible, but of course this cannot be easily achieved. Sorting will almost invariably use a standard sort program, supplied by a software house or manufacturer. The use of the first and last pass own coding option should always be borne in mind. In extreme cases this can save a whole program. If the mechanics of the manufacturer's sort routines are fully understood, it may be possible to arrange that the data to be sorted is produced by the preceding program in such a way that the first pass of the sort is made easier and faster. Similarly, if the first pass of the sort is process-limited, a reduction in storage used by this pass in forming strings may bring the pass into balance. In a critical situation, the pass in phase two of the sort might be saved by reducing record size and hence increasing string length out of the first pass.

Procedure segmentation

The segmentation of the total procedure is the crucial stage in designing the computer procedures. Generally, the smaller the number of different programs or runs the better. This is, however, an over-simplification. It is usually necessary to segment procedures because:

- one program would be too large to be accommodated in the store available;
- the complement of peripherals is inadequate;
- necessity for sorting involves a split in procedures.

Combination of procedures reduces set-up time and may enable one or more passes of a file to be saved. Segmentation into separate runs will tend to increase set-up time but can be used to reduce demand on peripherals. In particular cases of processor-bound runs, it may be possible to effect a reduction in processing time by partitioning into two balanced programs or by moving a particular processing function into the next program.

File design must be examined carefully to ensure that the right degree of compactness has been achieved. If one or more programs handling a given file are file-limited, it may be worth attempting to reduce the physical size of the individual records, so reducing file-passing time. Such compacting will tend, as a secondary effect, to increase the processing time, since extra coding will have to be inserted, into the programs concerned, to unpack the compacted fields. Conversely, where runs are computer-limited, it may be worth expanding the records, so easing the task of unpacking the record fields in an attempt to obtain a balance.

Where the system is to be run on a multi-programming machine, it is highly desirable to avoid computer-limited runs since this will tend to limit the flexibility of this facility. By combining operations for individual files, it is possible to reduce passing of sequential files and also reduce the number of accesses to a random file.

If several files can be consolidated onto one specific peripheral, fewer peripheral devices are used and control of operations can be automatic. The reduction in buffer areas will save main store and there will be a further reduction in duplication of information; but this method does give rise to more complex programming.

Where a system is split into a large number of subsystems or modules, each perhaps with its own files to process or maintain, it becomes essential to incorporate into the system a means of transferring data from one to the other. This could include:

- input data;
- data generated by one program for use by another, such as invoice records, produced by an invoicing program, which are to be used by the statement program;
- feedback, such as outstanding orders to be entered on a second run, or communication to the next run as a result of an action on the previous run;
- error signals, such as actions to be suspended because of an error situation, where errors found in handling data in one program affect another;
- file maintenance data, where one program in the system is responsible for maintaining the files for all other programs.

To perform any of these functions there will need to be, for batch processing systems, some kind of communication transactions, and for real-time systems, some area of storage, either main or backing where the necessary communication information can be held.

Finally, the amount of processing in any particular run will be determined by:

- the amount of main store available;
- the peripherals (including storage devices) available;
- the sequence of files and sorting requirements;

- the restart and recovery facilities required;
- the use of program utilities;
- the degree of expansion/flexibility to be built into the program;
- the program's ability to detect and deal with any errors which arise;
- the timing constraints.

Documentation

The output from computer procedure design will be, above all, a program suite specification which will provide the programmers with a description of the computer functions of the system. It is described in Chapter 22; it is emphasised that the extent to which the systems analyst specifies programs depends on the individual installation. The systems analyst must however document the computer procedures to provide as a minimum:

- a complete and unambiguous statement of the steps required to produce specified computer outputs;
- a means of testing the completeness of the stated requirements before detailed programming work commences;
- a means of communication between the systems analyst and programmer;
- a record of work done;
- a basis for project control.

TIMING OF COMPUTER PROCEDURES

At various stages in computer procedure design the systems analyst will need to estimate the time required by the computer procedures (both internal program time and external input/output handling time). This is necessary for the production of an effective design, for justification of the system, and for scheduling the live operation of the system.

Timing of computer procedures is a complex subject that depends not only on the machine speed, but also on the machine configuration. For this reason, it can be covered only in general terms in a book of this nature.

When timing computer procedures, essential factors to take into account are:

- set-up times;
- processor times;
- peripheral times;

- simultaneity;
- external procedures.

Set-up time

Set-up time is the amount of time that is required by the operator to:

- load programs;
- prepare peripherals;
- carry out any other actions required.

Because of the human element involved, it is difficult to work out precise timings. These will often depend on the layout of the machine configuration and the sequence of the operations.

If programs are stored on magnetic tape or disk and the program library file is always on-line, the time taken to load programs is negligible and can be absorbed in the set up times given to other peripherals.

Some of the characteristics of the job may increase the set-up time, for example, it takes longer to line up complicated preprinted stationery than to load a simple blank set of stationery. Quite often, however, some of this time can be overlapped with other jobs by preloading.

As an example of typical set-up times, it might take one minute to set up a tape deck or disk drive or paper tape reader; it might take two minutes to change the stationery on the printer.

If there are several operators, each can be responsible for different peripherals. Although the nominal time may be high in terms of man-minutes, the total elapsed time is that of the slowest operator. Thus the overall time to set up an input validation program with, say, a paper tape reader, one magnetic tape, and the printer (for production of error reports) may take two minutes. To set up an update program having one master file, an input and output file, or a sort program using, say, three work tapes, the set-up time may be three minutes.

Processor time

The type of central processor will affect instruction times and the number of instructions to be executed.

In order to time the internal processing of a computer run, it is necessary to determine:

- the average instruction time for the particular kind of job;
- the number of instructions to be executed.

All manufacturers supply average instruction times for their machines. These are weighted according to the incidence of arithmetic instructions in the program, as these types of instructions often take appreciably longer than non-arithmetic instructions. The number of instructions in a program varies with its complexity.

A program will normally comprise an initialisation routine, one or more loops, and an end routine. The initialisation routine and the end routine will be performed once only, whereas each transaction will have to pass through one or more of the loops. Some transactions may appear infrequently and, however complex, their effect on total processing time will be negligible. The timing calculation therefore is:

Number of transactions × number of their loop instructions
× average instruction time + time for initialisation and end routines.

Accuracy in estimating can be improved with experience by:

- recording the estimates;
- counting the number of instructions in the main loop of existing programs;
- asking experienced people, programmers or systems analysts 'Is this estimate reasonable?';
- comparing the estimate and subsequent operating performance, with the finished programs and investigating any significant difference.

It must be emphasised that this method of estimating processor time, or any other, can never be completely accurate but for all practical purposes it can be adequate. Where there are several loops, and transactions can take one of several paths depending upon the type of transaction, it is sufficient to decide which paths most of the transactions take and then time the program on the basis that all the transactions go through this loop.

Peripheral time

Card reader

The card reader is unlikely to achieve the rated speed unless it is double buffered. In addition, card readers usually have a mechanical clutch between the motor and the card movement rollers. Sometimes this has three contact position, so that if one position is missed, it has to wait for one third of a revolution, for the next point. One third of a revolution lost on each cycle will reduce the performance from 1,000 cards per minute to about 660 cards per minute. The next reduction will be to 330 cards per minute. When reading cards, the whole card has to be read regardless of the number of columns punched. (With a card punch the speed actually achieved will depend on the amount of data punched on each card and not the total space available.)

Paper tape reader

Paper tape reader performance must take account of the number of characters to be read which includes delete characters, control characters and blank characters in inter-block gaps. Although it is desirable to have gaps to differentiate between transactions they should be kept as small as possible. Winding and unwinding must be allowed for in set-up times.

Line printer

The three factors which affect printing time are:

- printing speed;
- spacing speed;
- skipping speed.

Spacing is faster than printing, and skipping is faster than spacing; the difference lies in the number of lines.

Various times are taken for building up a line of print, moving a document forward one line to prevent over-printing, moving a document forward two lines to leave a space of one line between printed lines, and the time skipping over eight or more lines.

To time printing the following factors must be known:

- number of documents to be printed;
- number of lines to be printed on each document, including blank lines spaced;
- number of lines to be skipped on each document.

Total print time can be determined by the following formula:

> Number of lines printed or spaced divided by print speed plus number of lines skipped divided by skip speed.

The speed of a chain printer is affected by the character set used and this must also be taken into account when determining the print speed.

Document readers

The speed of document readers is limited by:

- size of paper;

- amount of data on each page;
- the form of the output.

If the document reader is operated off-line with output to paper tape or punched cards, its speed will probably be limited by the speed of the punch. If the reader is on-line to the computer, speeds will depend on the physical rate of movement of the document.

Magnetic tape

Before timing a magnetic tape transfer the following must be known:

- tape speed, in inches per second;
- recording density, in bits per inch (bpi) but relating to characters (cpi) or bytes (bpi);
- number of tracks;
- start/stop time in milliseconds (ms).

The nominal rate of transfer in terms of characters per second is derived from the tape speed, the recording density and the number of tracks. Tape speeds are fixed for a given deck, but a choice of recording density is often available. Transfer rates are quoted in 1000 characters or bytes per second (eg 40 kc/s or 40 kb/s) and normally these are the maximum speeds.

Actual transfer rates will always be less than nominal rates because allowance has to be made for inter-block gaps.

Before a block of information can be read from magnetic tape, the tape has to reach full speed. After the block of information has been read, the tape will slow down and stop. Thus the larger the block size, the smaller the number of inter-block gaps and the fewer the number of start/stop operations that have to take place, producing a faster effective rate of transfer.

For a specific magnetic tape deck, it is useful to derive a graph showing the relationship between block size and actual rate of transfer. This will reduce the amount of calculation involved in timing magnetic tape transfers; the formula reduces to:

Total number of characters transferred
divided by actual rate of transfer

A further timing consideration is the time taken to re-wind a full reel, where a file is multi-reel. Re-wind times vary with manufacturer and could be between one and four minutes for a 2400 foot tape.

In the following example, a file contains 2000 records, each of 1000 characters unblocked, held on a magnetic tape driven by a 60 kc/s tape deck with a stop/start time of 10 ms;

$$\text{The total data transfer time} = \frac{2000 \times 1000}{60{,}000} + \frac{2000 \times 10}{1000}$$

$$= 33 + 20 = 53 \text{ seconds}$$

$$\text{The effective transfer rate} = \frac{2000 \times 1000}{53}$$

$$= 38 \text{ kc/s}$$

If the same data is blocked in ten-record blocks:

$$\text{The total data transfer time} = 33 + \frac{200 \times 10}{1000} = 35 \text{ secs}$$

$$\text{The effective transfer rate} = \frac{2000 \times 1000}{35} = 57 \text{ kc/s}$$

Magnetic drums

The two factors which affect timing on drums are:

- rate of transfer;
- rotational delay.

Before transferring any block of information we have to wait until the start of the block is underneath the read head. On average, this takes half a revolution. The formula, therefore, for timing drums is:

> Total characters transferred, divided by the rate of transfer, plus the rotational delay.

This formula is correct for reading data. When writing data it may be considered necessary to read back what has been written to ensure that it is recorded correctly. If normal hardware accuracy is not considered to be adequate, then allowance must be made for one extra revolution for each write operation.

Magnetic disks

Disks suffer from the same rotational delay as drums, but there is an additional factor, the amount of time it takes for the head to move to the appropriate track.

For any disk unit, the manufacturer will quote minimum and maximum head movement times. The minimum time is the time taken to move from one track to the next. The maximum time is the time taken to move from the innermost to the outermost track. Time taken to move intermediate distances does not, necessarily, have a linear relationship with the number of tracks moved.

There are many factors which affect the overall time spent in transferring data from disks. For example, the time taken to find any given record will depend, first of all, on the method of addressing, whether by indexes or direct, and then upon whether the record is at the first address given or overflow areas have to be searched. Thus, the file structure must be taken into account. Nevertheless, it is possible, using the formula quoted above, to calculate the overall length of time to find any specific record.

A commonly used disk pack consisting of six 14-inch diameter disks (10 recording surfaces) takes approximately one minute to load on to the disk drive. Typically, there is a maximum capacity for the disk pack:

- each track contains about 4000 characters;
- each disk surface contains 200 tracks, say, 800,000 characters;
- each disk pack contains 10 surfaces, say, 8,000,000 characters.

Although an average time to locate and input one block of data to main storage is 100 ms, this may be halved or doubled as a result of good or bad file organisation. The timing considerations for inputting one block of data are:

- arm movement time, which depends on the number of tracks the head crosses to reach the required track from the previously used track. The head moves in either direction along a fixed radius of the disk; it does not return to the disk's centre nor circumference during movement and it remains positioned at a selected track until another track is selected;
- head selection or switching time, which is usually negligible;
- rotational delay (or latency) being on average the time of half a disk revolution – say 10 ms;
- transfer time for one block of data is the number of characters in the block divided by the transfer rate, typically, 200 kc/s (notice that an entire block is transferred although only one record in the block is

required). The transfer rate depends to a large extent on the speed of rotation of the disk, eg 20 ms per revolution of 4000 characters yields 200 kc/s. Thus, the transfer time can be expressed in terms of disk revolutions, eg half a revolution to transfer a block of 2000 characters being half a track capacity.

The timing considerations for outputting one block of data from main storage to the disk pack are: arm movement, latency and transfer times (as described above), plus an allowance for checking that the block has been recorded correctly. The block is recorded with a 'cyclic check', which can be considered as a foolproof parity system, but, unlike magnetic tape recording, the block is not automatically input and checked for accuracy. Instead, an optional 'write verify' facility is available to the user which does re-input and check the block. The effect of this facility, usually taken for master files, is to increase the output time, by the time for one revolution of the disk, say 20 milliseconds, which is necessary to reposition the block under the read/write head and to re-input it.

Finally, the average time taken to input a block of say, 2000 characters, to amend it in main storage and to output it back to its original position (assuming that its position is known to the program) is given by:

- one arm movement of say, 85 ms (assuming that the arm is not moved between input and output to transfer another block) for 100 cylinders;
- two half revolutions to position the block under the read/write head;
- one complete revolution to check the recording;
- two half revolutions for block transfer.

Thus, the total average disk time is 85 ms and three complete revolutions of 20 ms giving 145 ms in all, plus any central processor (CP) time (see S J Waters, *Introduction to Computer Systems Design* NCC Publications, 1974).

Simultaneity

Two further factors which will affect the run time of any given program are:

- simultaneity of peripheral operations;
- CP time used by executive and operating systems.

Peripheral operations

Simultaneous peripheral operation is permitted by hardware provided that the particular processor can run the necessary peripheral devices simultaneously, and that the degree of simultaneity possible is not limited by a particular peripheral control.

Control limitations upon simultaneity will depend on the number of channels available for peripheral transfers.

It is possible to have several card readers, punches and printers transferring at once on slow speed multiplexor channels. However, there is no simultaneity between transfer operations using the same control or the same channel, so where there are many tape decks on one channel, it is only possible to write to one tape at a time. It is, however, possible to overlap such things as:

- magnetic tape rewind;
- disk head movement;
- rotational delay.

Channels are small computers in their own right, with their own limited instruction set. They can have several input/output areas on one device, so that it is possible to read in record after record as a buffer against uneven processing time, although normally two areas are allowed per file. They can carry out a sequence of instructions without any interruption of the main processing, although it is not usual to do this.

Data is placed in or taken from main store in competition with the main processing program, which also requires the storage mechanisms. This is done by cycle stealing where the main program stops for one main storage cycle and then continues. For timing purposes, it is necessary to know:

- how wide is the data path;
- main store cycle time.

CPU operations

Sometimes calculation is only intermittent, but lengthy at that time. Whilst this is going on the peripheral units stop. The interrupt system services peripheral units when they demand attention and then permits processing to continue. It also makes demands on processing and storage, and all of this takes time.

With one job in a machine, and a basic type of operating system with no complications, say, 20 % processing time is the overhead. With several jobs in a machine and many units working with a sophisticated operating system, 50 % of processing time is the overhead (someone in the installation must be able to advise on this).

External procedures

In addition to being able to determine how much time is going to be spent on the computer, the systems analyst must also be able to tell the user when, during the day, week or month, reports will arrive. This has to be considered in the light of the computer operations and the company environment.

Major factors to be taken into account include the time spent in data preparation and other user procedures. The timing effect of these factors depends on various conditions:

- staff efficiency;
- working conditions;
- suitability and legibility of source documents;
- number and type of data capture and other ancillary machines available;
- method of data capture – whether cards, paper tape, magnetic tape, or documents.

In addition to the external procedures specific to the job, interactions with other jobs both clerical and computer must also be considered. The jobs which the systems analyst is designing will have to be slotted into the existing load. Only then can delays that could occur during production running be evaluated.

Bar charts are commonly used to plan and optimise the timing of data flows and loading of the system being developed. The effect of the data flows of other connecting systems or subsystems must also be considered. Although the systems analyst is not responsible for the day-to-day running of the operations, and user procedures of all the other systems and jobs, their possible adverse effect on the system being designed must be assessed and acceptable timing arrangements made.

The timing of external procedures can have a significant effect on the elapsed time to produce computer output; close attention to this factor can be more important to a particular system than fine timing of computer operations.

SUMMARY

The systems analyst has to specify for the programmer the computer procedures which the system requires. The various procedures can be defined using tools such as computer run charts, computer procedure charts, network charts, interactive system flowcharts, and decision tables. There will normally be four types of procedure – validation, sort, update (or process) and print. Update procedures will vary depending on the file organisation and access method of the master file, and can be further broken down into serial, direct and database processing. The design of computer procedures needs to take into account: facilities like buffering and blocking, special methods for handling large files, and program segmentation/combination considerations. Finally, the computer procedures must be evaluated for efficiency in terms of estimated timing. Computer procedure design is, therefore, closely related to all other aspects of computer subsystem design.

14 System Security

INTRODUCTION

Considerable publicity has been given in recent years to damage to computer equipment (eg, from fire and flooding) and computer-related crimes and this has made organisations more conscious of the need to make their computer systems secure. Most breaches of security, however, are not dramatic like these; rather they are minor accidents, errors and omissions which can accumulate into serious losses, if they occur regularly. The systems analyst has a responsibility to design systems which are secure against such events.

An organisation is 'secure' if neither its ability to attain its corporate objectives nor its ability to survive can be adversely affected by an unwanted event. An organisation may be secure against *some* threats, but insecure against others. No organisation can be 100% secure, however much planning is undertaken and however much money is spent on countermeasures: any organisation, or indeed any person, is liable to suffer from unwanted events.

Computer security is defined similarly. A computer-based system is a combination of many assets or resources (eg, hardware, software, people, data, procedures, data transmission facilities) designed to perform some function or to provide some service. A computer system is secure against a particular threat (eg fire) if countermeasures have been taken to reduce to an acceptably low level the amount of loss which the threat may be expected to cause over a given period of time (eg a year). There are three types of loss which an organisation does not want its computer system to suffer:

- loss of availability;
- loss of integrity (ie, accuracy);
- loss of confidentiality.

Some systems may not contain anything which is confidential, but the first and second types of loss are applicable to all computer systems. A loss of availability, integrity or confidentiality, whether accidental or deliberate, whether for a short or long time, will adversely affect the ability of the computer system to carry out its intended tasks.

In other words, a loss of availability, integrity or confidentiality is a breach of the security of a computer system; this, in turn, is a breach of the security of the organisation which uses the system.

A threat to a computer system is any event whose occurrence would adversely affect one or more of the assets (or resources) which make up the system. Figure 14.1 contains a list of the seven categories of assets and shows examples of assets which may be present in a particular computer system.

Each of these assets is threatened by one or more of the following unacceptable events:

- interruption;
- disclosure;
- modification;
- removal;
- destruction.

And these unacceptable events can happen accidentally or be deliberately contrived. Thus there is a large number of ways in which the security of a computer system is threatened. Examples of some of the possible breaches of security are:

- accidental interruption of communication;
- accidental destruction of hardware;
- accidental modification of software;
- deliberate removal of programs;
- deliberate disclosure of information.

There are three other categories of accidental threat:

- those involving the malfunction of some component (eg a hardware or software fault), as opposed to an accidental misuse of a component (which is a threat caused by a person);
- those which occur naturally and which damage or destroy equipment, buildings and other physical assets. Floods, hurricanes, some fires (those caused by lightning), snow, ice, earthquakes, earth tremors and so on belong to this category;

Category	Examples
Hardware	central processor, magnetic disk drive, terminals, modems, shredders, storage cabinets
Software	operating system, compilers, applications programs, audit routines, security dumps, back-up copies
Data and Media	master files, input data, output files, software documentation, operating procedure documentation
Communications	telephone and telegraph circuits, postal services, private data carrying services, exchanges
Environment	building structure and fittings, power supply, air-conditioning plant, cleaning services, catering facilities, lift services
Organisation	management structure, management policy, technical staff, administrative staff, secretarial staff
Support	maintenance staff, auditors, consultants, suppliers

Figure 14.1 Categories of asset in a computer-based system

– those which affect people and their capacity to do their normal work. Death, injury and disease belong to this category.

The systems analyst must examine the system being designed for all the possible breaches of security (ie all the events, both deliberate and accidental, which can adversely affect the assets of the system) and develop a strategy to prevent or minimise losses to the system as a result of any breach.

RISK MANAGEMENT

The major steps in the method known as the risk management approach for providing cost-effective security are:

- identification of assets;
- identification of threats;
- measurement of threats;
- identification and measurement of countermeasures (safeguards);
- selection of countermeasures for the risk control programme on a cost-effectiveness basis;
- implementation of the risk control programme;
- monitoring and review of the risk control programme.

Threats can and should be described quantitatively so that someone can decide how much protection each should receive.

For any threat, the exposure equation $E = N \times I$ expresses the exposure (E) of the threat (the expected value of the loss per year due to the threat) in pounds (£), dollars ($), etc. per year as the product of the expected frequency of occurrence (N) of the threat (the number of times per year that the threat is expected to occur) and the impact (I) of the threat (the expected value of the loss per occurrence of the threat).

The threats can be ranked in order of importance based on their exposure values. It is tempting to select a countermeasure for the threat which has the largest exposure value, and then to select a countermeasure for the threat which has the second largest exposure value and so on, until all threats are given protection or until the budget for countermeasures is completely allocated. This, however, is not the best approach. The aim should be to minimise the total expected cost of the sum of the losses which will occur despite the countermeasures. (It will not be worth installing countermeasures against some threats, and some countermeasures which are installed will only give partial protection against the corresponding threats.) It is also necessary to consider the cost of the countermeasures which are included in the computer security programme.

Over a period of time an organisation may suffer greater losses from the frequent occurrence of threats which individually cause relatively minor losses than from the occasional occurrence of threats which individually cause relatively large losses. A study by IBM has indicated that classes of threats may be arranged in the following order according to their frequency of occurrence:

- errors and omissions (over 50 % of all threats);
- dishonest employees;
- fire;
- disgruntled employees;
- water;
- intruders and other threats (under 5 % of all threats).

This order is supported by the findings of a survey by NCC of about 150 UK organisations; figure 14.2 shows clearly that errors and omissions occur much more frequently than other types of threat.

Type of Threat	Number of Organisations Reporting		
	No Breaches	Some Breaches	Significant Breaches
Errors			
Machine fault	15	121	16
Operator/Clerical error	11	132	15
Software error	24	123	12
Application program error	12	132	11
Communications error	57	84	7
Power supply/Air-conditioning fault	31	118	5
Other			
Fire/Flood	129	13	1
Malicious damage	140	2	0
Theft/Fraud/Unauthorised use	140	2	0

Figure 14.2 Breaches of security in UK organisations

It is important to realise that many threats (the number is unknown) occur and are detected, but are not reported (they are not disclosed to the public). Other threats (their number is also unknown) occur but are not detected, and so are also unreported. In the first case, the organisation may fear adverse publicity, and consequently loss of confidence in it by its customers.

THE PROTECTION OF A COMPUTER SYSTEM

There are four possible ways of handling a threat. One is to avoid it by altering the design or the specification of the system in some way. This may mean deleting some feature from the system or deferring the introduction of some feature until, for example, a new version of the operating system is released from the computer manufacturer.

Another method is to transfer the threat to another organisation by means of a contract or by insurance. Insurance will always have an important role to play in any risk control programme, including the computer security programme. However, insurance on its own, is unlikely to be the most cost-effective way of providing protection in many cases, should be regarded as a last resort.

A third method is threat retention. An organisation can carry many trivial threats itself. This means that the systems analyst (or a specialist on this problem) estimates what the loss will be and is prepared to bear that loss, as though it were a fixed operational overhead. This is referred to as active retention. If a threat is carried unwittingly, however, because it has not been identified, it is potentially extremely serious, since a large loss might suddenly occur without warning, without any contingency plan, and without any insurance cover. This is referred to as passive retention.

Finally there is threat reduction. This means that the value of the exposure (E) of the threat is reduced by a countermeasure, which has the effect of reducing the frequency of occurrence (N) of the threat, or the impact (I) of the threat, or both.

Specifying security measures

Security measures for a computer system should be specified at an early stage in the life cycle of the system, so that they can be planned in detail during the system design stage and the system as it is being developed. It is a management function to allocate responsibility for the specification of the security features that are to be included in the system. Many threats cannot be effectively countered unless they are handled in this way, because it is often uneconomic or impractical to add security features on to an operational system.

During system operation, management should allocate responsibility for the enforcement of security procedures. Each employee should know and understand what procedures each is required to comply with and what the penalties are for non-compliance.

As an example, responsibilities for security may be allocated for the following areas of a system:

- hardware;
- software;
- data and documentation;
- communications;
- environmental;
- personnel;
- physical access;
- system access;
- administrative procedures;
- the development and implementation of contingency and recovery plan;
- the testing and review (the auditing) of normal and stand-by modes of system operation.

An organisation should have a pattern or procedure to adhere to for the development of a system through all of its phases or stages. It should superimpose upon this a system security profile which shows the security activities which should take place and the security outputs which should be produced during each phase of the life cycle of the system. An example summary of such a procedure is shown (fig. 14.3).

A breach of computer security involves a loss of availability, integrity or confidentiality. It is particularly important to design the system with controls, both manual and programmed. The aim is to check the accuracy of the system and of the data which it uses and to report for remedial action any malfunctions, errors or inadequacies which are detected.

The systems analyst should at least be aware of physical security and personnel security considerations, even though someone else may be given the responsibility for these aspects of security. Physical security is concerned with the building where the computer is located and the supporting services, access control, fire prevention, fire fighting, flooding, and so on. Personnel security is concerned with staff recruitment, appraisal and termination procedures, training, career development, the organisation of jobs and the allocation of duties.

During development, a computer-based system should conform to standard procedures and be properly documented. This will reduce development time, and ensure that the documentation is complete and easy

PHASE NAME	PHASE I – PROJECT INITIALIZATION	PHASE II – INVESTIGATIVE STUDY	PHASE III – GENERALIZED SYSTEMS DESIGN	PHASE IV – DETAILED SYSTEMS DESIGN	PHASE V – IMPLEMENTATION PLANNING	PHASE VI – SYSTEMS IMPLEMENTATION	PHASE VII – POST IMPLEMENTATION EVALUATION
PHASE PURPOSE	DEVELOP Understanding of situation	SELECT Approach to problem solution	DEVELOP Preliminary data processing systems design	REFINE Preliminary data processing design into final systems design	DEVELOP Plans for implementation	CONVERT Proposed system into an operational system	DETERMINE if system and cost objectives have been achieved and benefits realized
SECURITY ACTIVITIES	ANALYZE Narrative description of project. DETERMINE if security measures will be required. DETERMINE . What data needs to be collected? . Who needs it? . Why do they need it? . When do they need it?	DEFINE Level of security required for system. DEFINE Criticality of system. . Vital . Important . Deferrable DEFINE Internal project security procedures . Security clearances for project personnel . Security procedures for project working papers and documentation	ANALYZE Systems inputs and outputs DETERMINE Confidentiality of data ANALYZE Tentative master file or data base requirements DEVELOP Tentative security profile for master file or data base ANALYZE Hardware, software and communications requirements DETERMINE Security software to be used	DETERMINE Specific Software procedures: . Software Linkages . File Labels . Programmer I.D.Codes DEVELOP Detailed Programming Procedures . Data Checks . CPU Model Checks . Program Life UPDATE Tentative Master File Security Profile DEVELOP Report Distribution Procedures DEVELOP Documentation Distribution Procedures DEVELOP File Backup Procedures.	PREPARE Systems security test plan: . Unauthorized Access to System . Invalid Data . Recovery from Backup PREPARE Operational Security Plan	CONDUCT Systems security test: . Unauthorized Access to System . Software Routines . Operating Security	EVALUATE Systems security: . Security Software Reports . Control and Edit Reports . User-Written Audits
SECURITY OUTPUTS	Analysis of need for system	Confidentiality and Criticality of System Project Security Procedures . Security Clearances . "Need to know" Classification . Disposition and Control of Project . Working Papers and Documentation	Confidentiality of Data Master File Security Profile Security Software Designation	Software Procedures Programming Procedures Final Security Profiles Report Distribution Procedures Documentation Distribution Procedures File Backup Procedures . File Retention Periods . Backup Frequency . Number of Versions Retained . Disposition After Release	Systems Security Test Plan Operational Security Plan	Security Software Audit Reports Security Evaluation Report	Systems Security Evaluation

Figure 14.3 Security activities and outputs during the life cycle of a computer-based system

to use and amend. System and program testing should be carried out adequately and potential security exposures during testing reduced to a minimum. Acceptance and authorisation procedures should be observed at certain defined stages during system development. System and program amendments should be authorised, documented, tested, controlled and disseminated. System audit procedures should be followed to identify attempted violations, measure the use of the system, control the security tables and audit the use of files.

SYSTEM DESIGN CONSIDERATIONS

The controls which are built into a system must not only detect the occurrence of an error but also locate it and provide some means of recovery from it; where necessary they should also restrict access to particular data and users.

Errors

The principal sources of error in computer systems are hardware failure on a file handling device which destroys a file physically or prevents access to it, software errors, operator mistakes, invalid data and program errors causing corruption of files.

Hardware failure

Transient hardware errors on peripheral devices are normally handled by the operating system. For example, an unsuccessful input/output operation may be retried, or brief operator intervention may be requested. These conditions do not persist after recovery measures have been taken. The errors which persist after recovery measures have been exhausted are considered to be permanent errors.

Permanent errors fall within the scope of program level and system level error recovery. Recovery from a permanent input/output error depends, to some extent, on the hardware considerations. In addition, recovery also depends upon systems constraints imposed at system design time which are not hardware-dependent.

Software errors

These are unpredictable. If they could be predicted, they would not be permitted to exist. Prevailing conditions largely determine to what extent a particular software error is acceptable.

Operator mistakes

No manual procedure is entirely reliable and for that reason systems should always be designed to minimise operator intervention. To minimise errors in any necessary intervention, the action required from the operator should be standardised as far as possible.

Invalid data

Errors in primary input data should be detected by the validation procedure as described in chapter 12.

Under no circumstances should errors in data be allowed to cause unscheduled interruptions in the operation of a system. Such an interruption would, therefore, be deemed a program error, caused either by faulty programming or inadequate system design. Errors can, of course, occur in system control data, such as job control parameters; where software checks are inadequate to detect these at the earliest possible time, the system should check.

Program errors

These may be caused by the systems analyst or the programmer but the systems analyst should anticipate possible errors and design the system with as little ambiguity and as much clarity as possible. The system should detect errors as early as possible in the operating cycle.

Data preparation and capture

Controls for data preparation and capture should be established as near the point of data capture as possible. Data input to a system must be accurate, and the integrity of data held in the system must be maintained. Inaccurate records may make it difficult for a person to obtain credit or employment, or may create ill-will between an organisation and its customers, so that they transfer their business to another organisation.

A number of points should be borne in mind:

- input forms, as they are completed, should be checked for content, accuracy, etc, and should always be authorised/authenticated before they enter the data preparation or data entry procedures;
- precautions should be taken to prevent the deliberate entry of corrupt data or the accidental loss of genuine data (eg use of serial numbers on input documents);
- controls should be established for batches of input documents (or individual ones) which can be reconciled with computer produced controls (chapter 12) and batches should be sequenced;
- inaccurate input data should be corrected and re-input within a well-defined timescale and subject to rigorous controls;
- control and validation procedures should be equally strict for normal processing and for emergency processing;

- access to blank input forms should be restricted; completed input forms should have a stated retention period;
- when input data is fed into the computer it should be fully validated (eg for illegal characters, range checks, missing fields, internal consistency, check digits, and control totals) as described in chapter 12. If data is rejected, new controls may need to be created for subsequent processing.

Master file protection and reconstruction

The procedures for preventing data loss or corruption and providing for reconstruction of data in the event of a loss constitute a most important aspect of security.

File protection

This is provided by interacting hardware, software and system factors. Protection provided by hardware includes:

- device interlocks;
- file mask facility on direct access devices;
- parity checking;
- the use of write/permit rings.

Device interlocks prevent any interruption or termination of input or output, once commenced, and contribute towards prevention of corruption of data by any hardware or operator error.

A file mask can be set for any direct access file, specifying hardware rejection of specific sets of commands:

- all seek instructions, thus preventing inadvertent head movement;
- all write instructions, thus preventing corruption of an input file;
- all subsequent attempts to set the file mask, thus protecting the integrity of the hardware protection system.

Software protection includes comprehensive label checking. On magnetic tape the file label refers to one file only, and software will check that the correct file has been opened. A similar procedure takes place on disks, except that the software will generally keep the labels for files on that unit in a particular disk location. Direct access labels usually incorporate complex controls for read/write or read only, where multiple programs are to access a file. All file labels will generally have purge or expiry dates before which a file is not permitted to be over-written.

File reconstruction

This involves different approaches to the two principal types of file organisation:

- sequential organisation, including serial files;
- random organisation, including indexed sequential files which are not processed strictly sequentially.

With sequential organisation the problem is simple, because at any point of time during the processing of a master file, one can consider the file as consisting of two distinct parts:

- one portion completely processed;
- the remainder completely unprocessed.

If this point of division is identified before abandoning the run of a program for any reason, the program can restart at this point. By periodically noting this point of alignment, using a check-point facility throughout the processing of a file, any section of the processing can be rerun in isolation.

With random updating of a file on a direct access device, the problem is quite different. Whilst the logical classification of records into updated and non-updated still exists, the file can no longer be divided into two distinct physical areas corresponding to these. Furthermore, as the processing of a single record may be spread out in stages over the entire course of updating, the sharp distinction of updated and non-updated records becomes blurred.

It may be necessary to devise a system whereby each individual act of updating is a separate restart point. A unique serial number could be recorded on each record identifying and corresponding to each updating action. A further complication is that a file might be accessed by more than one program concurrently. If the sequence of updating is significant, the fact must be faced that due to the complex interaction of hardware and software, it will not be possible to reconstruct exactly the same sequence by rerunning the programs. One solution to this problem is to write all updating records, or copies of updated records, to a log file in the sequence in which they are applied. The processing can then be repeated by a program which reads this instead of the original input data.

Back-up

Back-up for the hardware, the data and the system is totally bound up with file reconstruction. It refers to the support available to an installation for use when one or more pieces of equipment necessary for the normal operation of any system become inoperable for any significant period of time. The importance of considering back-up in systems design cannot be stressed too strongly. However reliable any equipment may be it is bound to fail some time and although the average frequency of such failure can be predicted, the individual occurrence cannot. It is the systems designer's responsibility to ensure that the ensuing disruption of service is minimised.

Hardware back-up

This can best be illustrated by considering a magnetic tape installation. It is often advisable to have at least one tape deck more than is required for the running of any system. Either one deck more than the maximum requirement should be provided or all applications should be designed to run with one deck less than the total complement. In the event of failure of one tape deck, processing can continue.

When a critical piece of equipment becomes inoperable, such as a fixed disk or a central processor, the systems analyst and the user must establish a procedure to cope with this occurrence. It may consist of negotiation to have the system run on another machine with the same facilities, or duplicating facilities on site. However simple the procedure is, it must be established *before* a failure occurs.

Data back-up

This represents insurance against actual loss of data due to equipment malfunction, program failure or accident. The type of data back-up feasible depends largely on the size of master files.

The example (fig. 14.4) shows a simple brought-forward, carried-forward updating system, with separate files. The well-tried reconstruction method used here is to retain the input data file and the old brought forward version of the master file. By retaining the input, for two or more cycles (the 'grandfather-father-son' system) the latest version of the master file can be reconstructed at any time. The more cycles retained, the higher the degree of security.

The value of this must be offset against the additional cost of the storage medium and the off-line storage space. On direct access media it is essential to use separate disk packs for the different versions. Otherwise, a single accident could wipe out the total reserve of data back-up. Often, on direct access media, the updating system necessitates the master file being updated in situ.

The parent file is destroyed by the action of updating, but a copy of the parent file can be retained. The frequency of making this copy will depend on the file-size, the length of reconstruction run acceptable, and the degree of security required. For example, if a file is copied before every six runs of the updating program, the cost of producing the copy will be about one

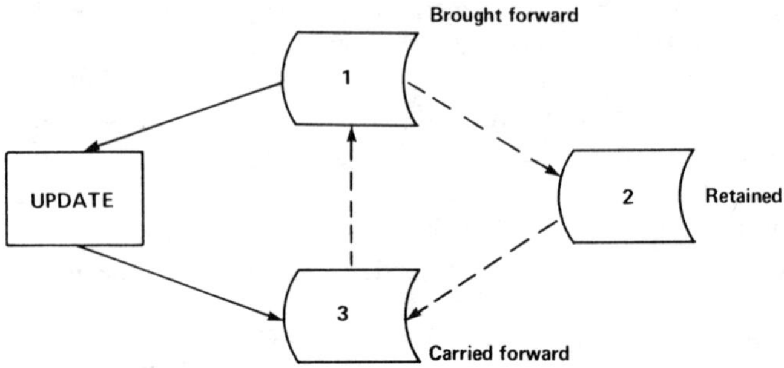

Cycle	Son	Father	Grandfather
1	1	2	3
2	3	1	2
3	2	3	1

Figure 14.4 Cycling of files for data backup

sixth of the cost of copying it every run, but the length of a reconstruction run, if necessary, will be six times as great. With really large files, it may well be impracticable to copy the whole file and various approaches may be made. For example, it may be satisfactory to copy a portion of the file, prior to each run, covering the whole file in a cycle, or to make an initial back-up copy of the file and then subsequently keep a copy of each record updated. Periodically, the updated records can be merged with the master copy. Copying the main file usually gives an ideal opportunity to perform file tidying, reorganisation or restructuring. It is also a good opportunity to carry out total reconciliation procedures, which cannot be done during random processing.

System back-up

A system maintaining a large file, on a fixed disk, may have a requirement to continue operating, even when the fixed disk is inoperable. If so, an alternative processing scheme must be designed which will permit the system to continue functioning.

At the worst, the bypass system shown in figure 14.5 could allow the system to continue to limp along, handling only the more important processing. Alternatively, it might be possible for the bypass system to handle the full work load, at the expense of other work. The example illustrates back-up, bypass and reconstruction procedures on magnetic tape of a random file updated normally on a fixed disk.

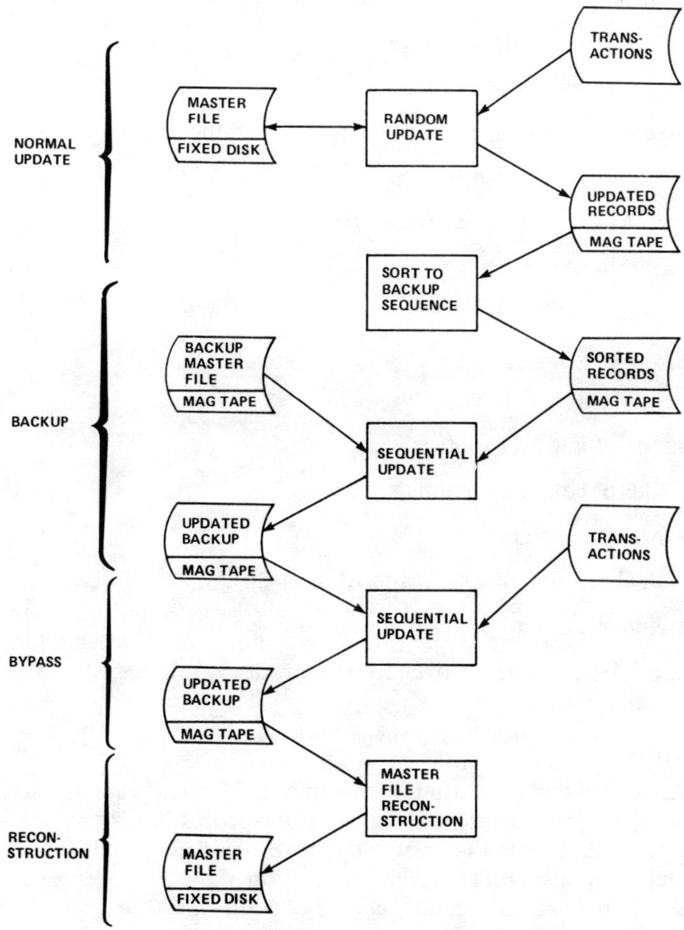

Figure 14.5 Backup, bypass and reconstruction procedures

In the worst situation the system may require back-up clerical procedures to ensure continued operation in the event of a prolonged breakdown.

Checklist of back-up copies

Back-up copies are required ideally for the following:

- system documentation;
- program documentation;
- operating procedures documentation;
- program source deck or tapes;
- program object deck or tapes;
- job control language cards or tapes for production programs;
- table card deck or tapes;
- operating system tapes or disk packs;
- master files;
- transaction files;
- report files;
- form masters;
- supply of key pre-printed forms;
- master of company manuals;
- printer carriage tapes (if used);
- control panel wiring diagrams (if used);
- documentation of duties of key personnel;
- description of the hardware configuration including all peripherals and all options.

File controls

The file security and processing checks imposed by hardware and software go a long way towards ensuring accurate processing. It is necessary, however, to ascertain that not only have files been read and written accurately but also that all the transaction data has been accurately processed. There are two main methods of controlling this.

One is to recreate the control totals established on input as a check against the input control totals. These may be:

- quantity or value totals;
- hash totals;
- item counts.

Very often the total produced on the output report will serve as checks on processing.

A somewhat more complex approach, although based on normal accounting methods, is to establish a control account for the file and post all items as transactions to this account.

The account balance at the commencement of the run, plus or minus the transaction data totals, should equal the closing balance at the end of the run. The control accounts would normally occupy a separate block at the end of the file and may be incorporated in the end of file sentinel.

Certain types of file can be regarded as permanent files, but nevertheless, a procedure for amendment must exist. In this case practical security features must be built into the system, rather than machine-based security features. For example, it is necessary to exercise considerable care regarding changes to rates of pay, selling prices, etc. A print out of amended items should normally be provided whenever an amendment is carried out. Filing of initiating documents must be subject to strict control.

Control against misuse

A file consists of many differing types of information and it is necessary to consider how confidential information, such as, the overall state of trading of a given commercial company or information relating to criminal convictions can be kept secure. Two main areas of security must be considered:

- the prevention of access to stored data without permission, ie the exclusion of invalid enquiries;
- prevention of unauthorised alteration, such as, a salesman entering incorrect statistics into a system so as to receive, fraudulently, commissions not due, or the alteration of a credit limit. Theoretically, a computer professional with a good knowledge of a credit control system could acquire considerable credit, well beyond the normal ceiling, unless means other than a stored credit ceiling were used for credit control.

There needs to be close control of the particular tapes or disks, restriction of handling to operators qualified and authorised to handle them. In certain instances, particularly with on-line systems, it may be advantageous to remove the most confidential data to manual media, stored in locked safes.

Control can be exercised at two points:

- access to the system;
- access to individual files.

Access to the system

The first security measure is to predefine all actions which are allowed at particular terminals; it is useful also to be able to identify the channel or line used by the terminal. Then appropriate measures should be provided for access control, eg passwords, hardware locks, badge readers.

Ideally all those passwords which entitle a user to access the system should be maintained within the minds of qualified users and within the computer, not on paper. Passwords should be either not printed/displayed or immediately obliterated once they have served their purpose.

Associated hardware devices, such as badges to be read, codes stored on an answer back drum, or hardware keys of the car ignition type which can be associated with certain terminals also need close control; this is a supervisory problem.

Access can be restricted to certain users, certain terminals, groups of users or groups of terminals, or a combination of these. Passwords should be only known by staff who need to know them and should be periodically changed; if a password falls into the hands of an unauthorised user it should be possible to delete it from the system immediately and to monitor any attempts to use it. Similarly each attempt to gain access to the system by a particular terminal should be monitored, logged and analysed.

Where confidential information is being entered into a system, unauthorised people should not have sight of it, and so the terminal data preparation device may need to be situated in a room with restricted access.

Where confidential data is being transmitted along telephone lines, it may be necessary to use cryptographic or enciphering techniques for security against wire tapping, misrouteing, and unauthorised access. These can be implemented either by hardware or software. The frequency of cipher changes is determined by the risks associated with the application.

The public switching network cannot offer a very high security. Particular care should be taken with systems using this network. For example, if it is possible for a user on a time-sharing service to close down a line without logging off the system, someone can then dial into the system and get access to the user's files without any log-on or check procedures.

Access to files

Access to given files can be restricted, using software routines which compare the access authority of a given user trying to access a file with the authority required for access, stored within the leader or header of the file.

Such restrictive checks must be incorporated into the executive program, and store accesses routed via such a portion of the executive. In certain instances this can apply a high overhead to the processor time to execute a job.

Alternatively, certain systems can have special protection facilities built into the computer hardware to prevent or restrict access to certain pages of storage when any given program is running. In this case the hardware page address, which may be a real or 'dynamic' address, related to real addresses via a look-up table, is compared with the access authority required, certain addresses corresponding to certain required authority. Again the comparison between the hardware address and user authority is made by means of an executive routine. If a user attempts to violate a hardware protection device which is preventing access, an interrupt is raised in most systems, access is prevented, and in many cases the operator is also advised.

Database management systems (DBMS)

Database Management Systems present particular problems for various reasons, one of which is the fact that they are implemented by vendors or mainframe manufacturers and are shared software systems. Thus their mode of working is well known to anyone working elsewhere with the same implementation. Control is essentially in the hands of the Database Administrator (DBA), or a staff member charged with that extra responsibility, who is the key figure in terms of both the efficiency and security of the database. Moreover, the increased integration of files creates additional problems.

Management must be sure that the security controls imposed on users of DBMS are adequate for their purposes. The user should have no means of accessing passwords of other users. Authority of access must be agreed at the right level in the organisation. Random or periodic checks should be carried out on the database to ensure its integrity and security (eg audit-trail, accounting control). Security controls should be available and built into the operating procedures to guard against copying, theft, destruction, or browsing of security sensitive data files (eg file dump, restart and recovery, operator action to override tape/disk labels or passwords, demonstration of DBMS to new users – this should involve non-sensitive files only), or their despatch over a network to another computer.

Output

Security measures need to be applied to output just as much as input. Output must be clearly and legibly addressed to avoid mis-direction (sometimes using a code to conceal the recipient's identity); and where confidential data is involved, there should be clear definition of who is allowed to access various classes of output. Output validation checks should be carried out by program to ensure the credibility and integrity of the output data.

The cost of security measures

No general rules can be set out to govern the selection of data security procedures. As in all other things, the ultimate factor that determines the selection of a specific system is cost. The systems analyst must evaluate each method in terms of money, because each involves the use of computer time as well as additional off-line storage. Measures can be justified in terms of the amount of computer time that would be wasted if no recovery mechanism were included in the system; also by showing that these maximise the productive computer time available, as opposed to alternative methods; and finally, in terms of the service the computer installation provides for the organisation.

The cost of error recovery may be considered in two categories:

- immediate cost;
- continuing cost.

Immediate cost

Error recovery procedures require more designing, programming, coding, testing and documentation time. Sometimes the additional programming can be separated into independent programs and subroutines included in processing programs. The more these subroutines accomplish, the simpler and faster the independent programs will be. The converse is also generally true. In deciding which should bear the burden of the work, the systems analyst should investigate whether the subroutine could be packaged and used by all programs.

The more work each recovery subroutine performs, the more main storage will be required for each program accessing direct access files. This can affect the ability to multiprogram. Additional main storage requirements need to be balanced against the impact on the size of the independent recovery programs.

The additional processing required per record to effect error prevention techniques increases the total processing time per record. This increase must be balanced against the amount of processing time required for the independent program.

Each error recovery scheme should be evaluated in terms of its requirements for off-line storage of tapes or disks.

As magnetic tape is much cheaper per character stored than disk, a magnetic tape deck can often be justified in a configuration for the purpose of back-up copying alone.

Each error recovery scheme must be evaluated in terms of numbers and kind of additional devices required per program. For example, the use of a log tape will require the allocation of at least one additional tape deck to the program. Additional device requirements also affect the ability to multi-program and may even cause program restructuring because of lack of devices on-line.

Continuing cost

Any error-recovery scheme requires a percentage of the total computer resources. The systems analyst's function is to keep this percentage down to the minimum as well as to maximise the amount of processing per record. The length of the reconstruction cycle must be balanced against the length of the reconstruction process. The shorter the reconstruction cycle, the shorter will be the process of reconstruction. However, the shorter the reconstruction cycle, the more often it will have to be run.

The cost of maintaining several generations of reconstruction data needs to be evaluated against the cost of having to reconstruct data manually. The degree of data protection required will depend on:

- the type and amount of data to be protected;
- the extent to which data is vulnerable to destruction;
- the frequency of processing;
- the extent to which data is altered during processing.

SUMMARY

System security has grown in importance in the last few years largely because of the increased scale of computerisation and the development of systems in sensitive areas. But the major concerns of any security policy (ie avoidance of errors and recovery in case of breakdown) are not in any sense new; they are an essential part of any system. This chapter examines system security in its widest sense and discusses the problems of risk management. Many of these issues are the responsibility of senior management. The systems analyst must design systems which detect, locate and correct errors, whether they occur at input, in processing files or at output; which provide measures for the reconstruction of files; which include prior arrangements for back-up in case of failure; which control access to confidential data; and which do all these things economically. System security considerations are all pervasive in their effect on both computer and manual subsystems.

Part V
Physical Design of Manual Subsystem

The physical design of the manual subsystem is concerned primarily with the interface between the computer and the users rather than with a completely separate part of the total system. The orientation of the previous part has been towards the computer and its constraints on the system; the orientation of this part is towards the human being and the elements of the system which must be designed with users in mind.

Chapter 15, the first chapter in this part, concentrates on the design of forms, mainly from the viewpoint of forms which are used by clerical staff to collect input data, but also taking account of the systems designer's need to design computer output documents which are easily interpreted and used. The topics covered are the content, layout, make-up and printing of forms. Chapter 16 examines dialogue design for systems which involve direct user interaction with the computer in conversational mode. Chapter 17 describes various types of coding systems and suggests some principles of code design to attempt to ensure accuracy of coding and ease of code usage by humans. Finally, chapter 18 deals with the design of user procedures with particular attention to workflow, office layout, staffing and equipment, work measurement and error handling.

15 Forms Design

INTRODUCTION

Paper is the traditional (and still the most widely used) medium for communication from the computer to the user, though the video screen is rapidly gaining ground. Many of the principles which apply to the design of paper documents for the transmission and storage of information apply equally to the design of layouts for screen documents. The purpose of both is to convey or collect information quickly and completely. If correctly designed, with the right users in mind, the required information will be entered to allow interpretation by the intended recipients, with the least time and trouble.

A *good* format enables relevant information to be obtained, transmitted, interpreted, stored, and retrieved at minimum total cost. A *bad* format can impede or prevent the required information being obtained, or, when obtained, cause it to be misinterpreted or even not understood.

A form is 'any surface on which is to be entered information, the nature of which is determined by what is already on that surface'. The surface could be paper, a wall-board, or a piece of plastic. In some instances the surface could be wiped clean and used again; with a VDU screen not only the information but the format as well can be made to disappear once it has been used.

Forms design is often considered to be a low-level activity to be delegated to a typist or clerk. The defects in many forms can be attributed to this attitude. In fact, designing forms is a highly demanding activity.

A good form does not occur by chance; it requires complete fact-finding, careful design and rigorous testing. The designer of a form is often responsible for every aspect of it, from determining the purpose, content and layout to checking the final version from the printer or on the screen. The forms designer should not pass an untidy draft to a typist, without very clear instructions. If a draft is not clear to the printer, then it will not be clear to the typist. The result may reflect, not the designer's intention, but the interpretation of such by the typist working within the limitations of the typewriter.

Good form design is not possible without good system design: what is on the form cannot be considered in isolation from the purpose of the form and the contribution it is to make to the objectives of the organisation. The occasion for forms design may arise as part of a system study; or a request for a new form may lead to a study of the system.

The stages of form design must follow the stages of a project. There must be a definition of the objectives, comparison of present results with required objectives, specification of information requirements, and design of layout; then testing, education and training of user management and staff, followed by implementation. User involvement should be encouraged at each stage.

This chapter is concerned primarily with the detailed design considerations, but all the other stages are important.

Before starting to design a form, the systems analyst should be satisfied that the form serves one or several specific purposes which contribute to the objectives of the organisation; that there is no other form, either existing or proposed, which serves or can be made to serve these purposes; and that it cannot be combined with any other existing or proposed forms to serve the combined purposes with greater accuracy or at lower cost.

The forms designer has a number of decisions to make for each form. Some of these must be by agreement with the various users; others must be decided by the designer in consultation with equipment manufacturers or printers. Most of the decisions will be interdependent, calling for the analyst's judgement. The main considerations are:

- content
- layout

} common to all formats (on paper or screen);

- makeup
- printing
- paper

} specific to formats on paper.

CONTENT

The content of the form involves all the words, spaces, boxes, etc, that are to appear on the blank form, including the title of the form and any instructions for its use.

Title

The title should be brief and meaningful to the users, and normally the last word should be a noun (but not the word 'form'—there are several alternatives such as note, advice, list, report, record, analysis, application, notification, instruction, specification). If a form has a long title such as 'Advice of Goods ready for Inspection and Despatch', it is not surprising to find users substituting a shorter name. If the particular form was pink in colour, then it might be called 'The Pink', which of course would cause confusion with any other pink forms. Most forms include a stationery reference number to be used for ordering supplies. Care should be taken to design a short meaningful title which does not encourage the user to use the stationery reference instead.

Detailed headings

The words used for detailed headings will depend on the background of the typical users of the form, their level of intelligence, and whether or not the form is part of their job.

A literate adult speaker of English of normal intelligence when confronted with:

Surname

will write his or her surname. A semi-literate or a child might be totally confused, or write 'Yes'. If phrased as a question:

What is your surname?

there is a good chance that the person's surname will be entered, but not necessarily in that space. For such a person:

Write your surname in this space

is likely to produce the required result; but this may still not be good enough if the form is to be used as a punching document. A possible answer to this could be:

Please write your surname in the space below, starting in the left-most square and putting one letter in each square until you have finished.

Where the medium is paper, the same instructions have to be used for each user, and brevity is important to save space. Where the space can be reused, as on a VDU, the words of instruction can be varied to suit the knowledge and experience of different classes of user.

Headings should always be as brief as possible but meaning should not be sacrificed for the sake of brevity. The typical user's needs should be borne in mind, including both originators and subsequent interpreters of the information, and headings should always be tried out before they are included in the eventual form. Special care should be taken with headings on a multi-purpose form.

Instructions for completion

Ideally, item headings will be self explanatory, and, if separate instructions are required, great care should be taken to keep them clear and crisp (eg repeated 'if' clauses should be avoided). If possible, instructions should be seen at the same time as the heading or question (ie should not be on the back of the form). It is annoying to have to *search* for instructions.

Where an instruction is critical, such as 'USE CAPITAL LETTERS', it should be seen before the person fills in the entry and not following the entry or at the foot of the page. Where the instructions are numerous and detailed, they should be numbered in relation to the headings or questions.

On the whole, the designer should try to avoid using separate instructions. However, where they are necessary, they should be tried out on users in advance.

Pre-printed options

Where there is a limited and known set of possible replies, completion and interpretation can be speeded up and ambiguity avoided by asking for marks in one of a number of positions. This particularly applies with hand-held instruments for completion such as a pen, pencil or light-pen. Where only two replies are possible, then one can be deleted (though, if a typewriter is being used to complete the form the time taken is as long as for typing a full reply); where more than two replies are possible a positive indication is better, and boxes (up to a maximum of ten) can be offered for the user to tick. The boxes to be ticked must be placed close to the answers to which they refer, eg:

ROAD	☐
RAIL	☐
SEA	☐
AIR	☐

or:

ROAD	☐	SEA	☐
RAIL	☐	AIR	☐

but not:

ROAD ☐	RAIL ☐	SEA ☐	AIR ☐

This method is particularly useful for questionnaires, where a free-form reply would give a problem of interpretation before analysis. This does not dispose of the problem of interpretation, however; it transfers it to the person completing the form. It therefore imposes on the designer the responsibility for ensuring that the words used have the same meaning for all possible originators, and that the options are mutually exclusive and cover all normal answers. If there can be other possible answers, a space should be given for a free-form reply. It is important also to ensure that within any form there is a consistent approach to positive and negative replies.

This imposes the necessity for particularly thorough field-testing whenever the 'option' method is used. This makes it expensive, not to be used for short-life forms. (This is why, in Chapter 6, vol. 1, analysts are recommended to use the questionnaire method of fact-finding only if there is no practical alternative.)

If the completed forms are to be used as input to the computer, each box may need to be marked with a code (for instance, a column number). Any such code should be made as unobtrusive as possible to the person completing the form.

LAYOUT

Having determined the content of the form the next stage is to design an appropriate layout.

Direction of paper

Before planning any detail of the layout, it is necessary to decide whether to lay out the form vertically or horizontally. Factors to be taken into consideration are: amount of data, environment of the originator, equipment used, mailing and filing.

It is usually easier to handle and file paper with the long side vertical, and it is probably convenient to regard this as the standard. Exceptions are such things as ledger cards, designed to be filed upright, and usually requiring a number of columns, and any other form requiring either a number of columns or some particularly wide columns (an example is the Record Specification, Form S43).

If the form is to be mailed using a window envelope, the size and shape of envelope will dictate the position of the address panel and therefore, to some extent, the layout of the remainder of the form. If the form is to be completed by typewriter, accounting machine etc, the length of the platen may determine the maximum width of the form.

Title and references

Once the form has been completed, the identification which matters is not either the title or the stationery reference, but the filing reference. This should be in a prominent position. Where the form will be filed by its left-hand edge, the best position for the reference is the top right-hand corner. This leaves the top left-hand corner as the obvious place for the title. The position of the stationery reference is not important, so long as it is standard for all forms in the organisation and does not obtrude on the space required for the filing reference.

Thus, if the filing reference on an invoice is the invoice number, then the invoice number should be prominent and in the top right of the form.

Entry headings

As indicated above, it is important that printed words and headings be close to or within the space to which they refer. A common fault is to allow so much space for the heading that it is not clear whether the entry is to go in the same space as the heading or in a space alongside or beneath it (fig. 15.1a). If the heading is within the box, it should be placed as far as possible into the top left-hand corner (fig. 15.1b).

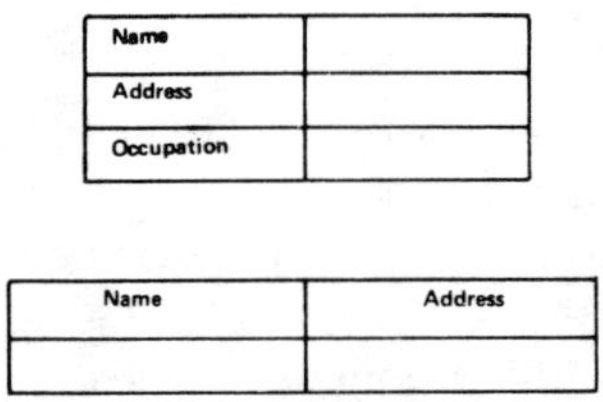

Figure 15.1a Too much space allowed for headings

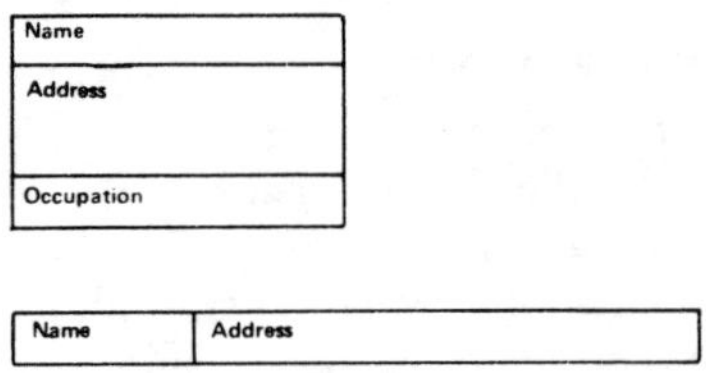

Figure 15.1b Preferable alternatives

For ease of understanding, related entries should be placed together and classified. This can reduce the amount of space taken up as well as improving understanding (fig. 15.2).

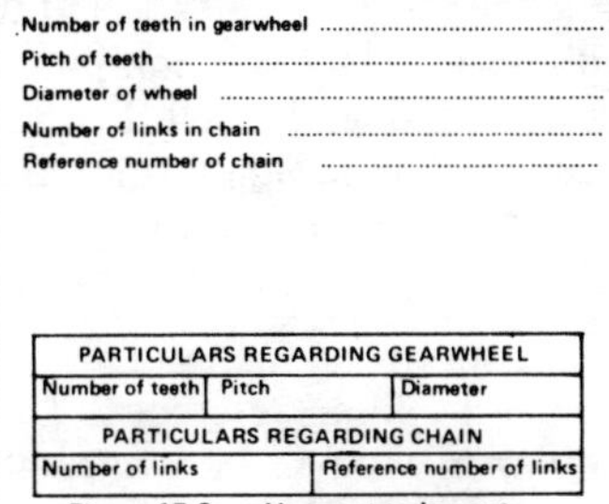

Figure 15.2a How not to lay out grouped headings

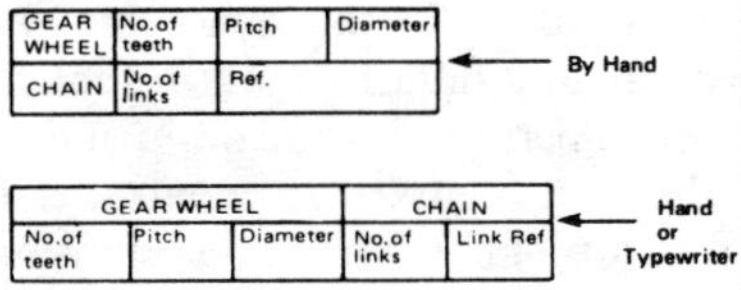

Figure 15.2b Preferable alternatives

The amount of space allowed for the entry should be determined by the number of characters in the entry, not by the number of characters in the heading. This applies particularly to column headings (fig. 15.3). The same principle holds good for tabulation produced as computer print-out.

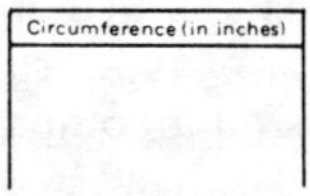

Figure 15.3a Column width determined by width of heading

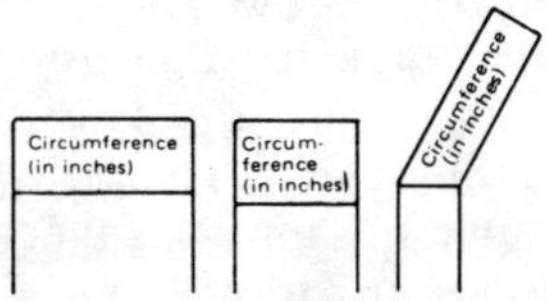

Figure 15.3b Preferable alternatives

If the equipment on which the form is to be completed is the kind which moves from left to right along one line at a time, such as a typewriter or teleprinter, then the headings should be above the entry space so that they can be seen and tabulating is kept to a minimum (fig. 15.3).

Entry sequence

It is not uncommon for the ideal sequence for the originator to be different from the ideal sequence for an interpreter, or even for different interpreters to require different sequences. It is essential to capture the data correctly: it is no good having a perfect layout for transcribing nonsense.

Where a compromise is necessary, particularly with a computer input document, account should be taken of the effect of entry sequence on time and accuracy of the originator, time and accuracy of the punch operator and computer time. The originator should not necessarily be expected to arrange data in the order in which it will be used by the computer: rearranging data is one job that computers do extrememly well. For the punch operator, the meaning of the different entries has no significance, but if there are related documents then it is worth taking some trouble to get the direction of eye travel consistent on them all.

Size and shape of entry spaces

A form giving a satisfactory overall appearance is likely to be a series of compromises, but an ideal size and shape should be decided on for each entry before any compromise begins.

The space allowed should be more than that required for the average entry, but for some exceptional entries it may be necessary to use a 'notes' space, the reverse of the form or an attached sheet.

The kind of data will determine the shape; if the entry is to be in words the horizontal dimension must be the greater, or if it consists of several sets of figures then the vertical dimension should be the greater.

For handwriting, the proportions of the shape allowed are more important than the actual dimensions. A normal handwritten character fits well in a space $\frac{1}{3}''$ high and $\frac{1}{4}''$ long; the $\frac{1}{3}''$ dimension is a particularly useful one where a form may be completed either by hand or by typewriter, since $\frac{1}{3}''$ gives double spacing on a typewriter (as well as on a computer line-printer). If space on a form is tight, and it is completed by hand by someone who normally works in a office, then the vertical dimension can be reduced to $\frac{1}{4}''$, but this is quite inadequate for the occasional user, who may have a blunt pencil, an unconventional writing surface or an uncomfortable writing position.

Typewriter spacing is standard at $\frac{1}{6}''$ vertically, but the horizontal spacing varies; Pica gives $\frac{1}{10}''$ and Elite gives $\frac{1}{12}''$. Where the machine to be used cannot be specified, $\frac{1}{10}''$ must be allowed.

Filing and gripping margins

It is not necessary for every form to have a margin drawn all the way round it; this only serves to decrease the space available for entries. If one edge is to be punched for filing, it should be left completely clear of information (though there is no reason why it should not be used for other purposes, such as title or entry instructions). A filing edge of $\frac{4}{5}''$ is adequate for use with most types of filing equipment. If the filing equipment is predictable, it will normally be an economy to have the holes prepunched as part of the printing job.

If the form is to be used with particular equipment, such as a typewriter or a duplicator, it is important to see that appropriate gripping edges are kept free of information. If it is to be used with particular input devices (eg OCR) then clearly the layout and especially the margins must be related to the special equipment requirements.

MAKE-UP

The term 'make-up' is used to denote all the physical features of a form apart from the printing and the paper. It includes the joining together of forms as sets, in pads or as continuous stationery, the provision by the printer of interleaved carbons or of any chemical coating for making copies, and any punching or perforating required. Making-up may well cost more than both paper and printing. A comparison needs to be made between the additional cost of make-up and the saving in user time, stationery wastage, or errors which would arise from not incurring the make-up cost.

If a number of different forms will be used by the same person from time to time, it can be an economy in time and in avoidance of waste, to have forms made up in pads of 50 or 100 with a strawboard back. If a fixed number of copies is required, then there will usually be a worthwhile saving of time for the originator in having supplies of the form made up into sets, joined together at a perforated stub, with either one-time carbon or a chemical coating.

If the form is to be completed by machine, then supplies should not be held in pads, but either loose or in sets. To avoid 'creep' in a typewriter, the joining of the sets should be at the top, or else continuous stationery should be used, with sprocket holes. However, continuous stationery is only worth using where it can remain in the machine for long periods, and only worth the expense of printing in large quantities; probably any quantity of less than 10,000 is uneconomic.

If a form is to be completed by hand, the number of parts in the set should not normally be more than three or four, otherwise the lower copies will be illegible. Some thought also needs to be given to the environment: it is no good having a flimsy multi-part set if it is to be completed by an outdoor worker. Where any kind of data preparation is to take place, it should be done from the top layer of a set, to minimise the possibility of error.

A chemical coating, such as NCR (no carbon required) is usually cheaper than one-time carbon for a two-part set, but more expensive for numbers greater then two in the set. A chemically coated paper needs careful handling, particularly after completion, since the surface remains sensitive to pressure, but there is the obvious advantage over one-time carbon that there are no carbons to remove before the copies can be used.

PRINTING

The printer should not be expected to guess at requirements; precise instructions need to be given, especially on critical dimensions, to enable him to realise the design, and he should produce proofs to allow the designer to approve the finished product before a large quantity is run off.

Single or double sided printing

The question of whether to print a form on both sides assumes that the amount of information required needs more than one side of the paper; there is no merit in spreading over two sides if one side is sufficient. Multi-sheet forms should be avoided if at all possible, since they cause additional problems in storing, requisitioning, feeding into any originating equipment, sorting, mailing, typing and retrieving. However, there are problems caused by double-sided forms, where immediate copies are required.

Double-sided forms are not possible with chemically coated copying papers. Loose carbons can be used, but this is time-consuming. There is also a limitation on the number of copies that can be produced by this method, since double-sided forms need to be of thicker paper than single-sided ones.

Serial numbering

The price of a form is only marginally increased by serial numbering, since this facility is available on standard printing machines. However, there is no reason for specifying serial numbering on every form. It adds to the problem of storing and issuing forms, since, to satisfy the purposes of serial numbering, there is normally a requirement to use the lower serial numbers first.

The advantage that serial numbering provides is security, particularly with documents such as cheques or invoices, which could be used as a means of obtaining money.

Use of lines

The essential purpose of lines on forms is to separate one area form another. On aesthetic grounds the use of many different line thicknesses is to be avoided; two thicknesses are usually adequate. A thick line can be useful to separate different groups of items, or to make individual items stand out, for instance, where selected items are to be punched. Where related entries are to be made by hand in a number of columns, faint guidelines are desirable, particularly from the point of view of later interpretation of the entries. For typing this is not necessary; horizontal lines are a nuisance to the typist, and there is no problem for the eye of the reader to follow a line of type across a page.

Type faces

The number of different type faces should also be kept to a minimum. A mixture of upper and lower case is to be preferred to all capitals, being easier to read. It is best to reserve capitals for use as headings for grouped items.

Type sizes

Sizes of type face are expressed in 'points', a point being approximately $\frac{1}{72}''$ vertically. This is not the size of the letter itself, but the distance between rows. Fig. 15.4 shows the point sizes most commonly available. The size selected must obviously depend on the space available, but other criteria are the age of the typical user – since eyesight deteriorates with age – and the lighting conditions in which the form will be used.

Colour

It may be useful for different copies of a form to be either of different coloured paper or printed with different coloured inks, though some colours can present problems with some copying processes. If there is no need for colour to aid recognition, it is probably best to keep to black ink on white paper. This avoids problems of matching different paper and ink stocks, and so minimises re-ordering problems as well as cost.

If a form is to be completed in black ink, the entries will stand out better if the headings are in a different colour, green being the colour recommended for best legibility. It is important that what stands out after completion is the entry rather than the heading, but this can usually be adequately catered for with black printing by position and type size of the heading. The exception is where the only printing required is a guide-line, as on line-printer stationery. For this, since the printing will often appear superimposed on a line, a faint coloured line is desirable.

Internal or external printing

Where the organisation is equipped with offset litho machines, there will be a useful economy in printing all straightforward forms internally, even if the artwork is done externally. Using this method, it is possible to get professional results but at a low cost of printing, and with the facility to print and store small quantities. It is usually uneconomic to use internal facilities for printing complex forms, particularly multi-part forms, for which specialist printers are properly equipped.

Order quantity

The quantity to order depends on rate of usage, expected rate of modification and cost of storage.

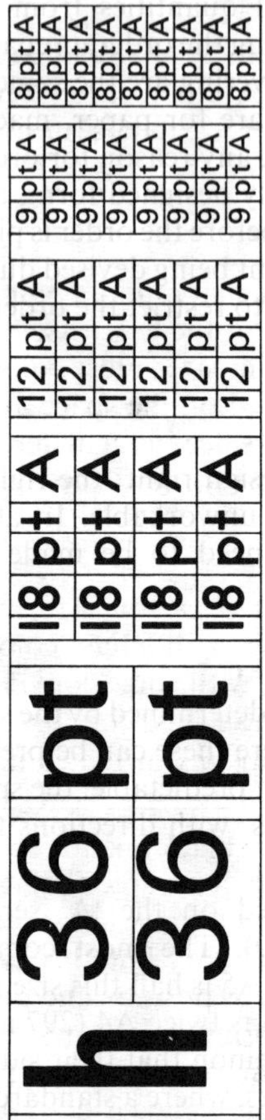

Figure 15.4 Point sizes

Buying stationery in small quantities from a commercial printer is expensive. His cost of processing the order, preparing the layout, setting up the printing machine and delivering are the same for 5,000 as they are for 500; the only different costs are for paper, machine time, handling and packing. This is not, however, any reason for ordering three years' supply of a form, particularly a newly designed form, at one printing. However carefully a new form is tested before the order is placed, there is a possibility of some suggested improvement being devised during the first few weeks of live usage. It is therefore wise to restrict the order for any new form to the lead time plus a few weeks.

PAPER

It is not enough to get the design right; the right design with the wrong paper may make the system unworkable. In addition to paper colour, mentioned above, decisions need to be made about size, weight and construction.

Size

The minimum size of a form is determined by the sum of the spaces required for the individual entries, where these can be pre-determined. If the space required for some entries is not predictable, the space allowed should cater for all but the rare exceptions, with directions for where any 'overflow' should be entered.

The normal sizes are based on the 'A' series of the International Standards Organisation (ISO). The most common size is A4, which measures 210 mm × 297 mm. A5 is half this size (148 mm × 210 mm) and so on through A6, A7 etc. A3 is twice A4 (297 mm × 420 mm). All these sizes have the property in common that their sides are in the proportions $1:\sqrt{2}$. There are also 'long' sizes, where a standard size is divided by 3 or by 4 along its longer dimension. These sizes are designated, for instance A4/3 and A4/4.

There is unlikely to be any saving in cost from specifying a size smaller than one of these standards. The printer may even impose a cutting charge, making the smaller size dearer then the corresponding standard size. There can be difficulties in handling very large or very small pieces of paper, and the practical size range is from A6 to A3.

Continuous stationery depths are measured in multiples of $\frac{1}{2}''$ because of the sprocket holes; widths are quoted metrically but are not fixed.

Weight

Paper is measured in grams per sq.metre (gsm). The most common weights are:

- 49 gsm: normally used for typing copy paper and suitable for most internal single-sided forms;
- 61 gsm: used for normal letterheads and suitable for any single-sided form not subject to long term repeated handling;
- 73 gsm: suitable for any double-sided printing and repeated handling;
- 96 gsm: standard for use with high-speed optical reading devices.

For use in carbon sets, 49 gsm is the maximum practical weight for all except the bottom copy.

Where large quantities of a form are used the cost difference between a lighter and heavier weight of paper may be significant. Even more significant may be the cost of postage, particularly if the forms are sent by airmail.

Construction

The internal construction of the paper determines its strength and opacity. For most purposes it is not critical, but there are exceptions, where paper is to be used in high-speed equipment, subjected to handling over a number of years or repeated impressions and erasures, or where translucency for copying is required.

Grain direction can also be important. If a card is to be filed vertically then the grain should run vertically; or if a form which becomes part of a volume is required to lie flat when the volume is opened, the grain should run in the same direction as the binding edge.

SUMMARY

Forms design is a highly demanding activity which is often the province of a specialist. It is concerned with ensuring that information is collected, interpreted, stored and retrieved with accuracy, clarity, ease and speed. The user's contact with the computer system is often via a form and it is crucial that the form does not impede that contact. The design of forms must be done in the context of the system as a whole and must take into account several interrelated factors: first of all, the content must be determined in the light of the system; an appropriate layout for the content must be designed; the physical make-up of the form must be defined; precise instructions for printing are required; and, finally, decisions have to be made about the size, weight and construction of the paper. The average medium-sized company is reckoned to use about 3,000 forms. Each of these is in itself expensive to design, produce and store; and yet the cost of using a form has been estimated at twenty times its cost of production. Clearly forms design is of crucial importance to system effectiveness and efficiency.

16 Dialogue Design

INTRODUCTION

Computer terminals provide a direct interface between a user and the computer, normally in the user's environment. The terminal may be used interactively to send messages which interrogate or modify the data held centrally, and receive responses which consist of a display of some part of that data; or to transmit batches of transactions to the central computer and receive relevant output (called 'remote batch working'); or to collect and store transactions for later transmission to the central computer for processing (called 'off-line working'). The design of terminal-based systems is concerned not only with input and output aspects of the computer system but also with the interfaces between the users, the terminal and the central computer.

The interactive use of terminals brings the power of the computer directly into the user's environment and, while this usually alleviates problems associated with accuracy and timeliness of data, it also creates problems for the user. Systems analysts should help users to come to terms with the extension of the computer to their own work areas. Methods of data control and quality control should be established. In a batch processing system these would be handled by human intervention in the punch room, data control section, or computer room. One of the major aids to effective and robust operation is careful design of the dialogue with which the user will be occupied.

DEFINING DIALOGUE OBJECTIVES

The first task is to identify which aspects of the user's requirements can be improved by use of a computer terminal. For example, the job of a booking clerk in a seat reservations office may be greatly improved by using a terminal to establish quickly the availability of seats, to provide answers to customers enquiries, and then to make firm booking of the seats chosen. Dialogues are usually required to fulfil two main objectives: the entry of data into the system and/or the provision of responses to enquiries.

Data entry

Terminals enable the user to enter data from the point where it is created into a central computer system. The system then checks the validity of the data as it is entered. A sales clerk, for example, can enter a sales order and, before the complete transaction is accepted, the account number, product number, quantity, etc, can be checked; if questionable entries are discovered they are sent to the clerk for correction (by referring to the source document) before the next order is dealt with; if the data on the source document is in error, then the transaction can be cancelled and the source document returned for correction to the originator (eg customer or salesman).

Enquiry

Direct access storage devices and suitable software facilitate the interrogation of files held in the central computer. In situations where fast and accurate responses are required, such as the seat reservation system mentioned above or a banking system where a customer's current balance is required at the time of cash withdrawal, this facility is essential. It would be possible to meet this requirement with a daily listing of balances, but the terminal offers greater speed and security and avoids the cost of daily printing and distribution of the listings.

More complex enquiry systems may be designed to satisfy a wide variety of search criteria, the precise nature of which is unpredictable; these are often known as information retrieval systems. For example, a sales manager may want information about which products will be affected by maintenance work on a particular production line and what action can be taken to overcome any deficiencies; this may involve analysing sales of the products by region, identification of alternative lines, or rescheduling of production on to other production lines. This kind of enquiry may be infrequent and not justify a specific program to be pre-written; in which case generalised information retrieval programs (or software package) can be used.

Data entry and enquiry

A combination of data entry and enquiry may be used in some situations where a transaction updates data held on main files. The seat reservation clerk initially makes an enquiry about the availability of seats when a potential customer wishes to make a reservation; when the customer makes a decision in the light of the response to the enquiry, a data entry transaction is made to update the seat reservation file.

DEFINING TERMINAL USERS

Having decided on the objectives of the dialogue, the analyst has next to establish some facts about the users of the terminals.

Dedicated and casual users

A terminal user may be described as 'dedicated' if the use of the terminal is a major part of the job; the 'casual' user is one for whom terminal usage is only a minor part of the job. A dedicated user needs a more coded dialogue with a wider range of options and has the opportunity to learn more quickly by experience on the job than a casual user. The latter may not be able to use the terminal efficiently if confronted by a complex dialogue; dialogues for those who rarely use a terminal must be self-explanatory and simple.

Active and passive users

An 'active' user is one who initiates the dialogue and leads the system through the main steps. A 'passive' user responds to messages sent by the computer which controls the dialogue and makes most of the decisions for the user on 'what to do next'. The seat reservation clerk who initiates enquiries and makes reservations is an 'active' user; a clerk, who is informed by terminal when errors occur and instructed as to the action to be taken, is a passive user. In general, messages for passive users need to be more self-explanatory than those for active users who know what to expect and what action to take. With dedicated, active users coded messages and specialised procedures can be built into dialogues, but this presupposes adequate training.

Intelligence of users

The level of intelligence of the potential users is impossible to assess and so the designer must try to relate the tasks to be performed to their general motivation and ability. Usually, of course, this will be dependent on training received, experience, and degree of usage. To cope with varying backgrounds, dialogues can be designed in an expansible way allowing for more detailed guidance where necessary. For example, messages can be included to ask whether a particular question has been understood, and a facility can be included to allow the user to enter 'Help' or '?'.

Persistent and intolerant users

Input messages to a terminal are often required in precise formats containing well-defined words and symbols. If an error is made in the input message, the program will need to take corrective action or to request a fresh input of the message. The latter option is easier to design and implement but depends on the characteristics of the user, in particular the willingness to persist at obtaining accuracy. A less tolerant operator might quickly tire of inputting marginally erroneous messages, in which case the dialogue should contain facilities for 'guessing' a correct input. For example, if a keyword is used to indicate a particular activity such as 'logging on' to the system, then the keyword could be accepted by the program either as 'LOG ON' or 'LOGON', in this case ignoring the blanks inserted. This could make life easier for the user.

TYPES OF DIALOGUE

A number of specific types of dialogue have been identified which are applicable in various situations.

Active dialogues

Dialogues in this category are generally of the enquiry type. The user initiates the input and analyses the responses (although active dialogues are often combined with other types of input). The main types of active dialogue are natural language, keywords, mnemonics and programming statements.

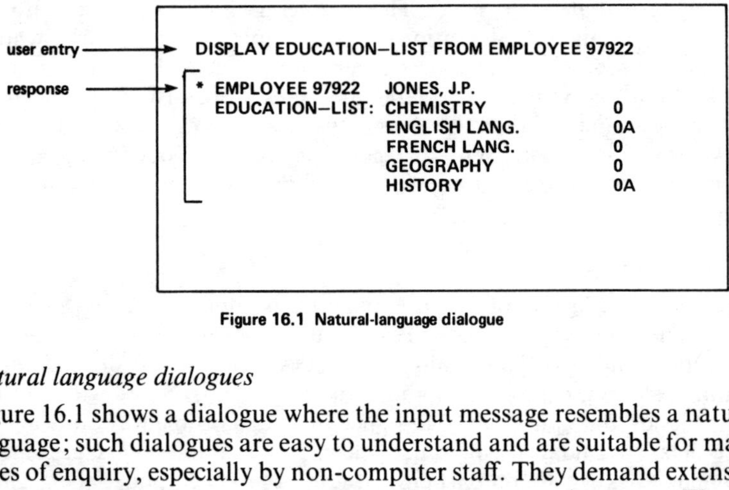

Figure 16.1 Natural-language dialogue

Natural language dialogues

Figure 16.1 shows a dialogue where the input message resembles a natural language; such dialogues are easy to understand and are suitable for many types of enquiry, especially by non-computer staff. They demand extensive interpretation by the control program and can be expensive to run. The use of natural language obviously requires caution to avoid misinterpretation of messages. The statement

FIND J. BLOGGS LTD AND CREDIT DETAILS,
IF OUTSTANDING-AMOUNT EXCEEDS CREDIT-RATING
THEN DELETE CREDIT-MARKER.

may mean that the credit-marker in J. BLOGGS' record must be deleted, or that the credit-marker in CREDIT-DETAILS record should be deleted.

Keywords

The use of keywords may retain some of the clarity of a natural language dialogue while removing the need for extensive interpretation. Figure 16.2 shows keywords in use, the keywords being separated from a value by an equals sign. Certain users may have difficulty in using keywords, as the statement COLOUR = RED may appear meaningless if one imagines that the letters of the word COLOUR are somehow made equal to those in RED. If keywords are employed then it should be possible to enter them in any sequence, otherwise even the most persistent user may become intolerant and give up.

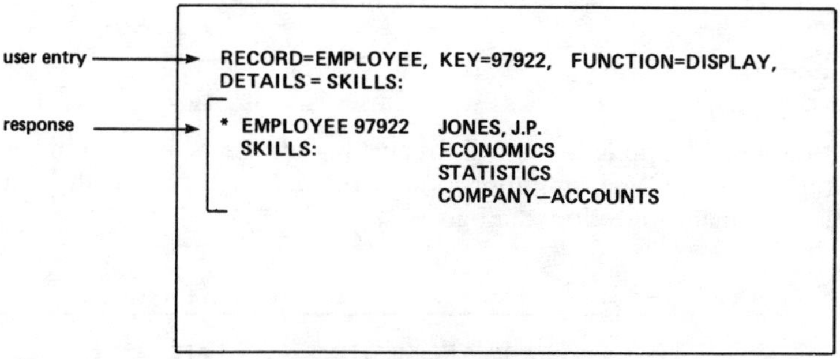

Figure 16.2 Keyword dialogue (enquiry type)

Mnemonics

Speed of input may be increased by using mnemonics in place of keywords. Mnemonics are short codes which have specific meaning. Consider the following entry:

N2F2

This is a portion of an airline reservation dialogue meaning 'I need to book two first class seats from the second of the flights shown above'. Note how much input would be required in a natural language version. With mnemonics, extensive operator training will be required, but experience has shown that well designed mnemonic dialogues provide fast and accurate service, providing the operator is dedicated, persistent, has a suitable environment and adequate training.

Programming statements

The dialogue in figure 16.3 uses typical programming statements. It is useful in situations where programming is involved, eg development, testing and debugging, but should be confined to users who have programming experience.

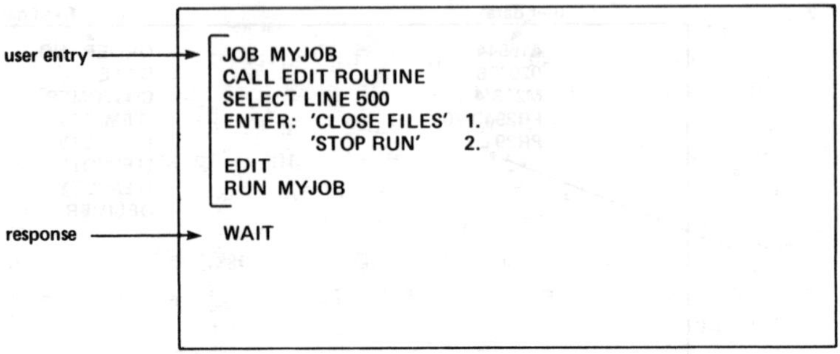

Figure 16.3 Programming statement dialogue

Passive dialogues

Most data entry applications will involve passive users and a number of different dialogue types are appropriate for such users, especially 'form-filling' and 'menu-selection' dialogues.

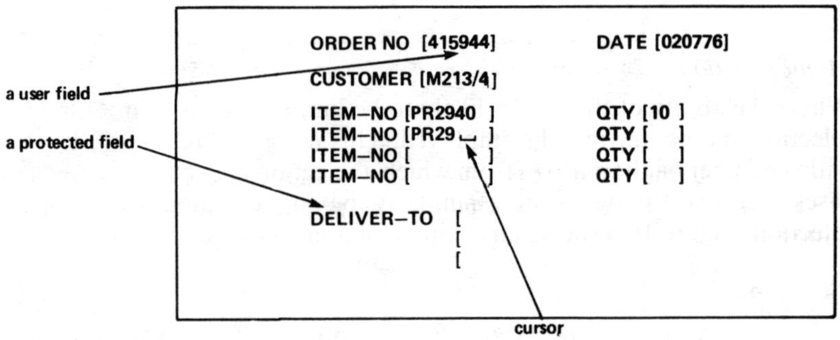

Figure 16.4 Data entry with form filling (formatted)

Form-filling

This is the type of dialogue in which an input document is simulated on a screen and the user fills in the 'form' by use of the input keyboard. Form filling dialogues may resemble very closely previous clerical methods of data entry, as in figure 16.4. These types of displayed 'forms' require extensive use of cursor positioning and, formulating characters in the control program. Certain fields in the display must be made indestructable by preventing any overwriting by the user. A more simple form from the design point of view is shown (fig. 16.5). Here, the name of the field is

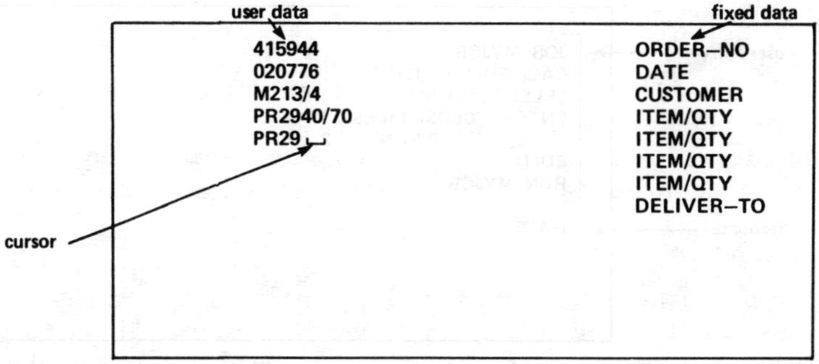

Figure 16.5 Data entry with simple form

entered at the end of each blank line, and if the data to be entered on a line is too long, then the fixed data may be overwritten. Form filling is a very common method of data entry dialogue, and it is particularly suitable where a transition from a similar clerical method is to be made, using the same staff.

Menu-selection

Where the ability of the user is likely to be limited then a form of 'menu-selection' data entry may be employed. This involves presenting the user with a number of alternatives from which a selection may be made. In some cases, enquiry dialogues may similarly be implemented using menu-selection. Figure 16.6 shows an example of a menu-selection dialogue.

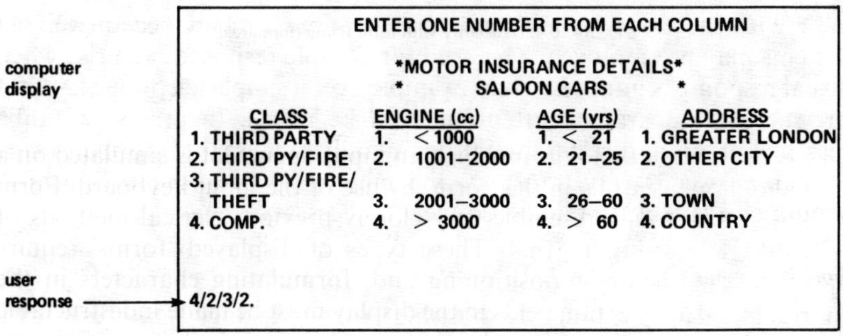

Figure 16.6 Menu-selection dialogue

Instruction and response

In situations where extremely low user-abilities are expected, such as when designing dialogues for use by members of the public, then a simple instruction and response system should be employed. Here, the terminal displays simple questions requiring simple answers, usually 'Yes' or 'No' and the computer establishes the data required from the answers.

Light pens

The use of light pens in a dialogue may speed up response to certain messages. It is likely that light pens will be most useful in menu-selection dialogues, especially where the user is not able to use a keyboard very well. With light pens, the user touches the screen display with the point of the pen at a place where the computer dialogue indicates alternatives, and the appropriate alternative is selected.

Special types of dialogue

Some dialogue types, though peculiar to particular applications, have features which may be useful in other situations.

Graphics and photographs

These normally require special devices with the terminals, and also imply a highly 'tailored' mode of use. Graphics terminals have large screens which can be used to display isometric or perspective views of complex structures. These views may be rotated, magnified and modified by the use of special function keys and light pens.

Photographs may be displayed on some devices, and are particularly useful in identifications systems, or in information archives, where pages of journals may be retrieved for viewing in microfilm form.

Voice answerback

Where it is necessary to give an audible response, then voice answerback systems may be employed. The need for audible response can arise where visual response is impossible, for example over a simple telephone handset, or where the operator's attention must be raised. In practice, audible response of this nature has limitations, mainly because the information in a spoken answer is very transient, and may easily be misheard. Facilities should be available to request a repeat if necessary.

Special devices

In specialised applications, it is possible to design dialogue features into the hardware itself. An example of this is a special keyboard for use with banking applications, where a large range of special function keys are made available, such as 'amount withdrawal' and 'balance enquiry'. Terminals of this nature which are designed specifically with a particular application in mind are economical only where there is a demand for large quantities.

It is possible to adapt general purpose terminals to specific applications by using various 'add-on' devices. One technique is to produce a template which fits over a keyboard so that keys which have been programmed for use in a special way are identified as such.

Buffered and intelligent terminals take over some of the input computing activity. For example, item separator symbols can be automatically inserted, data can be compressed or expanded or formatted; all of these reduce potential operator errors.

Error messages

The area where terminal operators need most assistance is the display of error conditions and, where possible, guidance on the appropriate correction procedures. Passive users require error-routine dialogues which gradually guide the operator back to the valid data condition. Active users may be given a more peremptory display such as

ERROR AT STEP 2.12

Both types of user require helpful, specific and quick error responses; but once the error has been displayed, much depends on the experience and training of the operator in deciding what action to take next.

Many errors in dialogues are attributable to format problems. Formats need to be defined precisely because the computer's ability to interpret entries is as yet limited. Format errors can be reduced by using variable length fields with field separators and by having a large number of small entry messages rather than a small number of large ones. But emphasis must also be placed on allowing the validation program some discretion in deciding the intended entry. One of the advantages of on-line data entry is the sophistication which can be built into validation procedures.

RESPONSE TIME

Response time in a real-time system is the elapsed time between the last character entered by the terminal user and the receipt of the first character of the response. There may be many different appropriate response times between different types of messages: response-time pattern in a dialogue is very much a part of the dialogue itself which must be designed to include the required response times.

Very fast response

Response times which are almost instantaneous have limited application. This is because their use can be confined to situations where the system requirements allow no 'thinking time' at all. A terminal user may feel 'coerced' if the machine responds almost instantaneously to each input message. Very fast responses are, however, useful where acknowledgement to correctly entered data is required, as this will speed up the data entry throughout. Also, if a badge reader is in acknowledgement of correct use, then insertion and acceptance of the badge or card may be very fast. In both these cases, the terminal itself may be capable of performing many of the checking functions required.

Responses of two seconds

Fast responses in the region of two seconds will be required in a variety of situations, such as when many rapid decisions must be made by the terminal user during the dialogue itself: when retrieving an item of data using logical deductions with several successive messages, the user relies on memory to achieve the correct sequence of inputs. Once the correct data has been found, the response time requirements may ease considerably. Other examples of fast response requirements are: searching through sequential records for a particular item, Computer Aided Design model manipulation, working with a light pen, and response to critical real-time messages upon which external actions depend.

These fast response requirements constrain system design, and should only be specified to meet genuine needs. At other points in a dialogue, longer and less constraining responses may be more suitable.

Responses of four seconds

The difference between two and four seconds may seem insignificant, but to a terminal user, successive responses which are too long will be frustrating and possibly disruptive. Responses of around four seconds are specified where there is no time-critical requirement, but where throughput would deteriorate if they were longer, eg error message responses, new display forms to be 'filed in', new graphic images to be examined, and non-critical real-time message responses involving external actions.

Responses up to fifteen seconds

These responses can be employed where a delay is acceptable, for example, on 'initialisation' or 'logging-on' to the system. Simple enquiries on a random basis where the costs of the system must be low, (eg a teletype installed in a warehouse for occasional stock enquiries) do not require a fast response. These responses may also be used at points in a dialogue where the user needs a 'breather', such as, for order entry, when an operator has completed an entry and is searching for the next order form.

It may be found that updating of files can be delayed until points are reached in a dialogue where a longer response is acceptable. This will help to obviate design constraints, as file updating can have a serious affect on response times.

Longer responses

If responses to inputs are longer than fifteen seconds, then interactive processing is likely to be unrealistic. Long delays will only be tolerated if the user can be released to do something else before returning to the terminal for the output response.

Implementing response times

Response times are an integral part of a computer terminal dialogue, and as such, systems must be designed so that the correct response can be achieved in each part of the various dialogues. Once a specification of the desired response pattern has been made, then the initial implementation of the system should conform to this, even though the number of terminals initially available is fewer than planned for finally. A response time which gets worse as more terminals are added to the systems can have an unfavourable effect on the terminal users: if necessary, a designed delay should be built in to simulate the ultimate response times. This may be modified as new terminals are added, until the final configuration and response times are reached.

THE TERMINAL ENVIRONMENT

All the effort involved in dialogue design may be wasted if the user is deterred by the environment from making good use of the dialogue. Concentration will be affected by telephones, people, ambient noise and other environmental intrusions.

Interruptions

Terminal users who suffer many unscheduled interruptions from time to time during operating periods make more errors and work more slowly than if otherwise. A dialogue which requires a reasonable level of concentration and application will be difficult to use if the operator is not permitted a fairly 'trouble free' working day. If it is found that interruptions, or a significant amount of simultaneous non-terminal activity exist in a particular user's work, then a dialogue with facilities for 'pausing' and 'recapping' may be indicated. Dialogues must be designed with the nature of working pressures in mind. Repetitive data entry dialogues such as form-filling work best and faster when the operator is not interrupted.

Dialogues for use in such situations should be as clear as possible to interpret visually, perhaps after attention has been drawn away from the subject in hand for a short period. It should be possible in such situations to be able to recall the state of a transaction so far. In highly distracting environments, some means may be necessary to attract an operator's attention, for example by the use of audio-signals such as bells or buzzers.

Comfort and convenience

The comfort of a terminal operator is an obvious, but occasionally neglected pre-requisite to efficient dialogues. Attention to detail from the point of view of ergonomics will help to ensure that the operator has enough room, can reach all the necessary keys and controls and can comfortably read the response messages. Terminal work stations are available with the required facilities for telephones, writing surfaces and filing systems. Levels of lighting are important with particular consideration given to the elimination of reflected light which may obscure part of a visual display.

Privacy

People usually prefer that others are not aware when difficulties are being experienced with a task. This is particularly true if it is also felt that they find the same task easy to perform. A terminal dialogue is an example of a task which may occasionally prove difficult to continue with correctly, and it must be expected that at times an operator will become 'stuck'. This condition will be more common with trainees, and with users who have little training or whose use of the terminal is casual. In anticipation of such situations, it is prudent to consider the aspect of privacy for the user of a terminal. By providing facilities where the user does not feel 'observed', the operator may approach difficult parts of a dialogue with more boldness, and thus be more likely to find out what went wrong. This may be done by physically siting terminals used by trainees or casual users in such a way that no windows from other rooms offer a view to the terminal's display. Another form of privacy may be provided by enabling the user to call in times of trouble for 'anonymous' assistance. This may be a 'debugging' portion of the dialogue which may be entered when necessary, or a supervising operator on a remote terminal who may be able to 'join-in' to resolve the problem.

DESIGN RULES

There are a number of points which can be made about dialogue design which summarise those made above:

- dialogues must be designed in such a way as to encourage the user and must be easy to learn;

- the system should be designed like a teaching machine guiding and supervising the user throughout the conversation;
- dialogues should reflect the user's view and terminology rather than the computer's or the system designer's;
- the user should never be left waiting and wondering; there must always be rapid feedback to any user entry, particularly in error situations;
- error messages should be carefully worded to guide the user in correcting the error; clarity of error display is of special importance;
- all messages should be consistent and logical in their appearance to the user;
- it must always be possible for the user to get help, ideally by asking questions in dialogue;
- thorough validation and editing routines should be built into interactive systems; source data is being entered directly without the human inspection procedures that are inherent in batch processing systems;
- dialogues should be designed to suit the user, the environment and the situation with the kind of flexibility that allows different speeds of operation for different user abilities and experience;
- the computer should be used to log the performance of different dialogues and to highlight difficulties experienced by users.

SUMMARY

The design of dialogues for interactive data entry or file interrogation, like the design of forms, can greatly affect both the efficiency of a system and user acceptance. The systems analyst must be quite clear about the dialogue objectives and the dialogue users before starting the design. With a picture of the user in mind, the analyst can then choose an appropriate dialogue type for active or passive conversations, and determine the level of response time which is required. The design of dialogues requires a sensitive approach to user needs.

17 Code Design

INTRODUCTION

The purpose of codes is to facilitate the identification and retrieval of items of information. The systems analyst will find code structures of various types being used by existing systems, but their structure will not always be the most suitable for efficient computer processing. Often existing codes are entrenched, not only in the system being changed, but also in other systems. Design of a new code requiring a great deal of time and effort may cause disruption in other parts of the organisation: in this situation, the existing code may have to be retained as the basis for the new system.

Normally a new code will only be designed when an entirely new computer system is being introduced; or when an existing code is outgrown and not capable of being extended; or where two organisations merge and require a single coding system.

This chapter provides a description of basic coding methods, including the advantages and disadvantages of each, to assist in the selection of the most appropriate code structure for a particular situation. Principles of code design are also included.

TYPES OF CODE

A code is an ordered collection of symbols designed to provide unique identification of an entity or attribute. To achieve unique identification there must be only one place where the identified entity or attribute can be entered in the code structure; conversely, there must be a place in the code for every thing that is to be identified. This 'mutually exclusive' feature must be built into any coding system.

Codes are also used for other purposes. Sometimes they are used in place of the name of an item; they can specify an object's physical or performance characteristics and they can be used to give operational instructions (eg which supplier, when to reorder, where to store, how long to keep) or other information (eg price, delivery date). They can also show interrelationships (eg between parts or customers), and may sometimes be used to achieve secrecy or confidentiality.

A common purpose is to classify objects for analysis. Within computer systems, codes are used for many of the above purposes, but also for economy of content and format; for identifying, accessing, sorting and matching records.

There are many possible coding structures. The main types of coding methods are discussed in this chapter (fig. 17.1). The codes discussed are 'pure' codes: many of the coding systems in real life are combinations of these basic types.

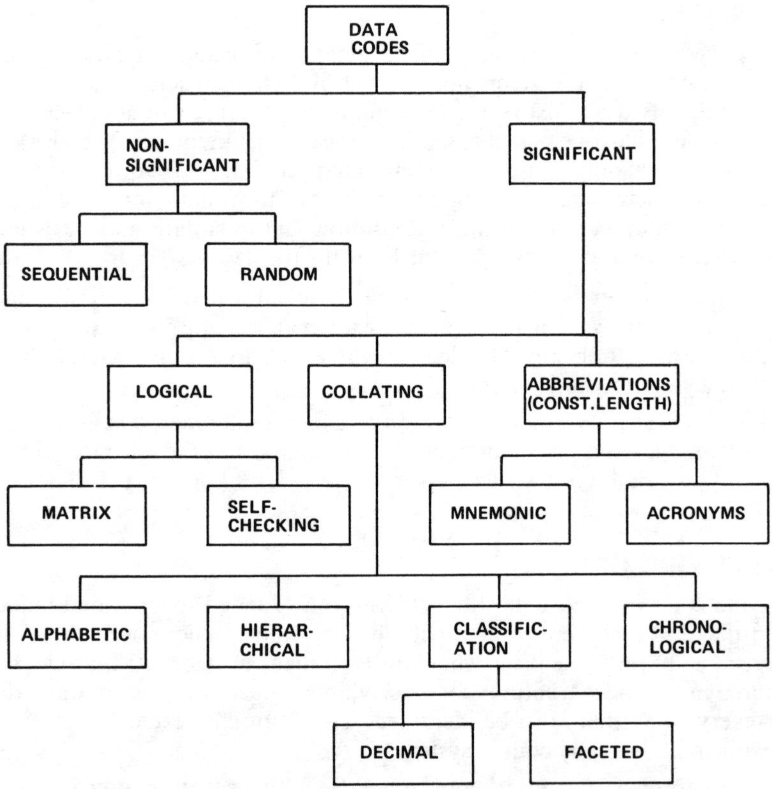

Figure 17.1 Forms of data codes

NON-SIGNIFICANT CODES

A non-significant code is one in which individual values are meaningless without some defined relationship to the entity or attribute which is being coded. Therefore, they are assigned only to provide unique identification. They can be considered to be more useful and robust for identification purposes than significant codes, which can become obsolete by change. The two most commonly used non-significant codes are the sequence number and the random number.

Sequential code

The sequential (serial or tag) number is simple to use and apply. This method of coding is merely the arbitrary assignment of consecutive numbers to a list of items as they occur, eg employee numbers might be assigned consecutively to employees as they are hired. The code value has no significance in itself but does uniquely identify the entity. It makes no provision for classifying groups of like items according to specific characteristics. It is practical when the only requirement is for a short, convenient, easily applied representation.

An advantage of the sequence code is that it can cover an unlimited number of items by using the fewest possible code digits. As new items occur they are simply assigned to the next-higher unused number in It is frequently used to give a unique reference number to entities (eg countries) which are composed of several elements identifiable in their own right (eg states, principalities, cities). With proper controls it is extremely useful in many applications and usually exists as a part of other more specialised coding systems.

Block codes

Block codes are a special type of sequential code in which sequences are grouped within sections or blocks as in the examples in figure 17.2. The Standard Industrial Classification is also an example of a Block Code.

Block coding systems of this kind are common. They allow expansion at the expense of initial extension. With practice, they can be useful for memorising the items or groups which they represent. All sections can be added or deleted easily. The user or a computer program can carry out simple checks on the code numbers as, say, orders are received or data is read by the computer. The code can also be used as a basis for sorting or for information retrieval. Block codes are often used where code numbers are issued at several different locations and the first digit indicates the location of issue.

Random code

The term random number is often applied erroneously to the sequential code just described. The difference between a sequential and a random code

01–10	Tea (¼ lb., India, China, etc.)
11–20	Coffee
21–30	Tea Bags, etc.

20,000 – 39,999	Weekly retail
40,000 – 49,999	Monthly retail
50,000 – 59,999	Weekly catering
60,000 onwards	Monthly catering

Figure 17.2 Examples of block codes

is the number list from which the code values are assigned. The random code is drawn from a number list which is not detectable in order or sequence. There are computer programs available to produce these random number lists. Each additional item to be coded is given the next number in the random list. In a sequential list, if 200 were the last number assigned, the next one will be 201. The next number in a random list might be 163. This method forces the coder to look up the next number on the list because there is no logical way to predict the next number.

This forced look-up is supposed to reduce errors in coding, but in actual usage it tends to introduce problems of control, and carefully used sequential lists have proved less error-prone than random lists.

SIGNIFICANT CODES

A significant code is one which, as well as providing unique identification, is designed to furnish additional meaning, which can yield logical, collating or mnemonic significance.

Logical codes

In a logical code the individual values are derived in conjunction with a consistent, well-defined logical rule or procedure (algorithm). Two examples of a logical code are the matrix code and the self-checking code.

Matrix code

This code is based on x-y co-ordinate locations or longitude-latitude co-ordinates. It is useful in coding two-component relationships. Code values can be formed by assigning the 'xy' co-ordinate numbers or by assigning sequence numbers. (The squares in the example are numbered both ways for illustration.) A code value is merely read from the appropriate square in the table when assigning code values to an entity. When decoding, the code value is located in the matrix and appropriate 'xy' attributes are obtained. Figure 17.3 gives an example of a matrix code; the numbers in parentheses are merely the matrix sequence numbers; the other numbers are the resulting code values.

x \ y	1 = Round	2 = Square	3 = Rect.	4 = Oval	5 = Irreg.
1 = Round	11 (01)	12 (05)	13 (09)	14 (13)	15 (17)
2 = Square	21 (02)	22 (06)	23 (10)	24 (14)	25 (18)
3 = Hex.	31 (03)	32 (07)	33 (11)	34 (15)	35 (19)
4 = Oct.	41 (04)	42 (08)	43 (12)	44 (16)	45 (20)

Figure 17.3 Matrix code

Self-checking codes

It is possible to append to a code an additional character which serves the purpose of checking the consistency from one point to another. This character, which is commonly called a *check character*, is derived by using some mathematical technique (algorithm) involving the characters in the base code. The check character feature provides the capability of detecting most clerical or recording errors. These errors are categorized in four types, ie transposition errors (1234 recorded as 1243), double transposition errors (1234 recorded as 1432), transcription errors (1234 recorded as 1235), and random errors (1234 recorded 2243) which are multiple combinations of transposition and transcription errors.

Several different techniques are employed to generate the check character. Each method has its advantages and disadvantages, according to the complexity or capability of the equipment involved in the system and the degree of reliability essential to the particular application.

The most useful techniques have been found to use the 'modulus' check character. With this method, each character in a code is multiplied by predetermined numbers (the weights); the check character is then the difference between the sum of these products and the next multiple of another predetermined number (the modulus).

To check a number having a check character, each character is multiplied by its weight, and the sum of those results is divided by the modulus; if the remainder is zero, the code is probably correct; if the remainder is anything else, the code is certainly wrong. The example in figure 17.4 shows the

Code 1 2 3 4 5 6

Weights: 1 7 3 1 7 3

Modulus: 10

Calculation of check character:

multiply each digit by its weight:

Digit		Weight		Result
1	x	1	=	1
2	x	7	=	14
3	x	3	=	9
4	x	1	=	4
5	x	7	=	35
6	x	3	=	18
Sum of results				81

Difference from next multiple of ten:

90 − 81 = 9

Check character = 9

Code with check character: 1 2 3 4 5 6 9

Checking:

multiply each digit by its weight (note that the check character has a weight of one)

Digit		Weight		Result
1	x	1	=	1
2	x	7	=	14
3	x	3	=	9
4	x	1	=	4
5	x	7	=	35
6	x	3	=	18
9	x	1	=	9
Sum :				90

Dividing by 10 leaves zero remainder, therefore code is probably correct.

Figure 17.4 Check character example

calculation for a check character; and in figure 17.5 the calculation for detecting errors.

Where codes are numeric, a numeric check character is normally used (called a check digit). Where codes are alphabetic or alphanumeric, a single

Suppose our correct code 1 2 3 4 5 6 9 was written down with the initial 1 changed to a 7, ie 7 2 3 4 5 6 9 (a 'substitution error')

Check

Digit		Weight		Result
7	x	1	=	7
2	x	7	=	14
3	x	3	=	9
4	x	1	=	4
5	x	7	=	35
6	x	3	=	18
9	x	1	=	9
Sum				96

Divided by ten the remainder is 6;

therefore an error has occurred.

Suppose the first two digits are reversed, ie 2 1 3 4 5 6 9 (a 'single transposition error')

Check

Digit		Weight		Result
2	x	1	=	2
1	x	7	=	7
3	x	3	=	9
4	x	1	=	4
5	x	7	=	35
6	x	3	=	18
9	x	1	=	9
Sum				84

Divided by ten the remainder is 4;

therefore an error has occurred.

This particular modulus 10 system detects 100% of single character substitutions and 88% of single character transpositions.

Figure 17.5 Detection of errors

digit cannot be used, for reasons which will become clear later, so either a single alphabetic or alphanumeric check character or two check digits are used. For the calculations, the alphabet follows the sequence of digits 0 to 9 with a = 10, b = 11 then .....z = 35.

Generally speaking, checking alphanumeric codes is a little clumsy. However, since an alphabetical code is shorter for the same capacity (5 alphanumeric = 8 numeric) the reduced length reduces error overall, despite certain offsetting increases in transcription errors. This latter increase is almost entirely made up of single transcription errors, eg 2 for z, especially where handwriting is involved. But the best modulus systems detect 100% of such errors. Thus an alphanumeric code with an alphanumeric check digit is a powerful tool, being cheaper to transcribe (through short length), and having a reduced error occurrence rate coupled with a very high error detection rate.

There are many variations of modulus systems in existence. The majority differ so slightly that any variation in performance is impossible to assess; others have hidden pitfalls. As a general rule, it is wise to use one of the widely accepted systems shown in figure 17.6, unless there are strong reasons to the contrary.

	Numeric Codes	Alphanumeric Codes
Low Protection	modulus 10 weights 7, 3, 1, 7, 3, 1 basic efficiency 90%	
Normal Protection	modulus 11 weights * basic efficiency 90.909%	modulus 37 weights * basic efficiency 97.297%
High Protection	modulus 97 weights * basic efficiency 98.969%	modulus 523 weights * basic efficiency 99.808%

* Oscillating arithmetical progression: weights ascend 1, 2, 3.....to a value just below half of the modulus, then descend from modulus minus 1 to just above half the modulus, repeating; eg, the weights for modulus 11 will be, from right to left, ...3, 2, 1, 6, 7, 8, 9, 10, 5, 4, 3, 2, 1

Figure 17.6 Commonly accepted modulus systems

Certain observations can be made about the choice of a modulus system. In general it is true that the higher the modulus, the higher the protection, provided that the following rules are observed.

- the modulus is equal to or greater than the base, ie the number of different characters that may appear in any position, (10 for numeric, 26 for alphabetic and 36 for alphanumeric);
- the modulus and the weights are 'co-prime', ie do not have any factors in common;
- repetitions of the same weight are separated by as many positions as possible;
- the modulus is preferably a prime number (10 is the only non-prime modulus in common use).

Overall, the proportion of invalid codes which escape detection is basically 1/m where 'm' is modulus; thus with modulus 11, 1/11 are undetected, which is to say, 91 % of errors which could occur will be detected, 9 % will not.

The application of the foregoing rules and the careful selection of the weights facilitates detection of the most commonly occurring errors and allows the remaining undetected errors to be reduced to those less likely to occur, such as random, or valid but incorrectly selected codes. In this way, the performance of modulus 11 may be enhanced to over 99 %. However, it can be dangerous to depart from the published details of the systems shown in figure 17.6, or to devise new systems, without a firm understanding of their mathematical basis.

For numeric codes, 10 is the minimum value of the modulus to satisfy rule 1, but it violates rule 4 (it is not a prime). Because of this, the choice of weights is limited, so in practice rule 3 has to be compromised. The result is a system with somewhat limited protection, but it is useful because it has none of the problems discussed below. For high protection, including 100 % cover of the commonest errors, modulus 11 is widely used. For very high protection, modulus 97 is good, but requires two check-characters.

For alphanumeric codes, 37 is the smallest prime number and therefore the minimum modulus, and is the obvious choice. If very high protection is required, then modulus 523, represented by two alpha-characters, is good.

For general protection, moduli 11 and 37 serve numeric and alphanumeric well, but share a common problem: how do you represent the two check-characters which are equal to 10 and 36 in the two systems respectively? Three solutions are adopted:

- omit codes yielding those two values of check characters; this is possible if the codes and their check characters are being newly issued, and is the best solution where practical. However if the check character is applied to existing codes, this option is not open;

- use another character to represent the problem check character value. Common choices are A or X for 10 with modulus 11 on numeric codes and * for 36 with modulus 37 on alphanumeric codes;
- use two characters; this can only be done where the check character is separated from the coded checks, or confusion will arise as to which position or weight applies to which character.

Where the weights have repeating groups, eg ... 7, 3, 1, 7, 3, 1, with modulus 10, and the code has similar length components, eg a 9-digit code divided into threes, as 769 243 561, the system will not detect any transposition of the groups, and alternative weights or modulus must be used.

At present, various national organisations in conjunction with ISO are studying a standard approach for check characters. If a standard is recommended it is likely to incorporate the five systems shown in figure 17.6. The main benefits of using a standard system are:

- easier checking of interchanged data, that is, where data is exchanged between different sites;
- standard hardware and software for generating and checking check characters should be more widely and cheaply available;
- better checking; the standard systems recommended will have been rigorously checked.

Collating codes

Collating codes are by far the most directly useful and the most frequently used. The collating code structure is designed so that when sorted by the code number, the items represented by the codes are placed in a predetermined sequence. This sequence is frequently the sequence of the output required from the computer for optimum use by people. Examples of collating codes are alphabetic, hierarchical, classification and chronological codes.

Alphabetic codes

For maximum effectiveness, alphabetical coding requires placement of all items in alphabetic sequence, then assignment of a code of ever-increasing value. Subsequent sorting of the code puts the items in the original alphabetical sequence, for example:

01 – apples;

02 – bananas;

03 – cherries;

04 – dates.

Normally, space is left between each item for future expansion. This code has some very strong points in its favour:

- ease of sorting into desirable output format;
- ease of maintenance;
- accessibility to the code list without initial encoding.

Unfortunately, this code has some disadvantages that can result in problems that are extremely expensive to correct. This is especially true in large, scattered systems where high rates of corrections or additions are necessary to maintain the list.

These disadvantages include:

- the necessity of coding the entire item list at one time to get reasonable spacing for new entries;
- crowding that requires renumbering to maintain sequence of new entries;
- relatively short life;
- the necessity of central control of number issues.

This code does, however, have a very useful place. Proper system design can utilize its good points and eliminate many of its shortcomings for certain applications.

In certain situations there is a requirement for alphabetic-derived codes, especially when dealing with indexes for large files of names and allowing entry to the file via the name index rather than a code number. Here, a fixed length code is desirable which can be derived from the name in accordance with certain rules (a type of logical code). For example the rules might be:

- use the first letter;
- ignore vowels and H, W and Y;
- replace the next three letters by:

  1 for B, F, P, V

  2 for C, G, J, K, Q, S, X, Z

  3 for D, T

  4 for L

  5 for M, N

  6 for R

  ignoring consecutive repeated characters;

– complete the code with zeroes if less than three digits are yielded on conversion.

Thus, examples of this code would be:

Smith	S530	O'Connor	O256
Smythe	S530	O'Connell	O254
Jones	J520	Connor	C560
James	J520	Lee	L000

This code is known as a SOUNDEX code because it is based on phonetic principles. Obviously it does not produce unique references but it groups together similar sounding names so as to ease searching for the specific name within each group. The grouping could obviously be made more refined by incorporating initials, sex, etc.

Hierarchical codes

The hierarchical code is a collating code which ranks entities or attributes by relative levels. It is very useful for many diverse applications. In its simplest expression, the hierarchical code arranges items in a predetermined sequence. The sequence may be increasing weight, length, diameter or other single attribute of the items.

As code requirements become more complex, pure hierarchical coding is seldom sufficient for large systems. New ways to create hierarchies have been developed using the basic technique in combinations with other codes. Hierarchical codes are still of great value in specialized applications or supplementary to a larger code system for indicating increasing values, organisation structures or levels of data summary control.

Classification codes

Classification is best described as the establishment of categories of entities, types and attributes in a way that brings like or similar items together according to predetermined relationships. A classification is by nature an ordered systematic structure.

The design of a classificatory structure must satisfy two basic requirements: comprehensiveness and mutual exclusiveness of its categories. Its scope must be broad enough to encompass all the items that need to be included in the various classes, and the definition of the classes must be exact enough to assure the existence of only one place for every item. Further, that place must be the same for every user of the classification.

Entities and types and attributes change continuously in a dynamic world. A viable classification system which contains them must be flexible enough to accommodate such changes. Its classes must be expandable. To be comprehensible, new and mutually exclusive classes may have to be added to the structure. Old classes may in addition have to be modifed or deleted.

Classification schemes are based on the viewpoint of particular people, called upon to do certain tasks at a specific point in time. As experience grows and circumstances change, the systems too must grow and change.

One of the most widely known classification codes is the Dewey Decimal System used primarily for indexing libraries or classifying written correspondence by subject matter. The following is a representative example:

300.	Sociology
400.	Philology
500.	Natural Science
510.	Mathematics
520.	Astronomy
530.	Physics
531.	Mechanics
531.1	Machines
531.11	Level and Balance
531.12	Wheel and Axle
531.13	Cord and Catenary
531.14	Pulley
531.141	Pulley, Compound

The decimal method of coding is designed to be used for identifying data in situations where the quantity of items to be coded cannot be limited to any specific anticipated volume. It is particularly well suited for classifying and filing abstracts of written material because it is able to handle an infinite number of items as they are added to any given classification.

Pure decimal code construction does not lend itself readily to mechanised data processing methods because fixed-code field definition is inconsistent with the decimal code expandability.

A number of devices may be used for machine processing of the decimal code, such as tagging variable length fields, special indentation and spacing, and blocked construction as in the following example.

CODE	*SUBJECT*
531000	Mechanics
531100	Machines
531110	Level and Balance
531120	Wheel and Axle
531130	Cord and Catenary
531140	Pulley
531141	Pulley, Compound

In this example, the decimal code has been converted to a six-digit fixed-field block classification code.

The organisation of the decimal code is retained, but the degree of expandability has been limited to ten subdivisions for each machine class. In most classification systems, classes are divided into subclasses, and subclasses are divided further into sub-subclasses. When coding these classes and subclasses, usually the code assigned is unique only within the subclass since the same codes are used to code members of another subclass. By example, the following illustration demonstrates the dependency of the identification of the class.

Class: Countries of the World

Members: Canada – coded CA
Sweden – coded SW

Subclass: First Order Administrative Division of Countries of the World

Canada (CA)

Alberta – coded 01
British Columbia – coded 02
Manitoba – coded 03

Sweden (SW)

Ålvsborgs Län – coded 01
Blekinge Län – coded 02
Gälveborgs Län – coded 03

In this example, the code 01 is a first order administrative division code representing two different entities (Alberta in Canada and Alvsborg in Sweden). In order to be unambiguous, the division code must be used with the country code as CA01 for Alberta and SW01 for Alvsborg. In this example, the division code is dependent upon the country code in order to yield unique identification.

When classified and coded in this way, the subclass or division code is a *dependent code*. When it is used with the country code, this collective code is also a *significant code*, because the code structure not only identifies the administrative division, but also the country to which it belongs.

The concept of dependency is not limited solely to classes and subclasses. For example, in certain applications different transactions are identified by a code consisting of parts which represent the organisation, the date of the transaction, and a serial number assigned to each transaction on that date. In this example, all three code segments must be employed to produce a unique transaction number. This too, is a dependent significant code of the composite data element named 'Transaction Number'.

Certain kinds of classification code are known as faceted codes. This kind of code represents different facets of the coded item. Each facet is given as many characters as required. For example a garment manufacturer's code might have as its facets:

	Suit	Man's	Size	Style	Material
	SU	M	38L	17	384
(or)	15	12	383	17	384

Chronological codes

As the name implies, a chronological code is assigned in the order of events so that each code has a higher value than the last code assigned. This is essentially the same approach as non-significant sequential. The difference is the attachment of time significance to the code number assignment.

Abbreviations

A frequent method of coding is by abbreviation of the name of an entity or attribute. The main ways of doing this are by mnenomic codes and acronyms.

Mnemonic codes

Mnemonic code construction is characterised by the use of either letters or numbers, or letter-and-number combinations, which describe the items coded, the combinations having been derived from descriptions of the items themselves. The combinations are designed to be an aid to memorising the codes and associating them with the items they represent. Mnemonic codes produce fewer errors than other types of code where the number of items is relatively small and stable. For example M and F are more reliable for Male and Female than 1 and 2, and Y and N for Yes and No.

Unit of measure codes are frequently mnemonic codes. For example:

cm – centimetre
m – metre
km – kilometre

Occasionally mnemonic codes are derived from names by removing vowels. For example:

Andrews	ANDRWS
Smith	SMTH
Tracy	TRCY

There are some problems connected with the use of mnemonic codes to identify long, unstable lists of items. Wherever item names beginning with the same letters are encountered, there may be a conflict of mnemonic use. To overcome this, the number of code characters is necessarily increased, thus increasing the likelihood that the combinations will be less memory-aiding for the code users. Also, since descriptions may vary widely, it is difficult to maintain a code organisation which conforms with a plan of classification.

Mnemonic codes are used to best advantage for identifying relatively short lists of items (generally 50 or fewer unless the list is quite stable), coded for manual processing where it is necessary that the items be recognised by their code. A common problem, however, is that the code is likely to be misapplied when specific code values are subject to change and users rely too heavily on memory. Thus, to be effectively coded with mnemonics, entity sets must be relatively small and stable.

Acronyms

The acronym is a particular type of mnemonic representation formed from the first letter or letters of several words. An acronym often becomes a word in itself. For example:

RADAR = RAdio Detecting And Ranging
HEW = Department of Health, Education & Welfare

Only when there are fixed lengths are acronyms considered data codes.

PRINCIPLES OF CODE DESIGN

The need to communicate with, and by means of, computers has made increasing demands on users to work with and understand computer codes instead of natural language. It must always be remembered that a code will be used by human beings, including people who do not have much familiarity with data processing. Codes should be designed with two features in mind: optimum human-oriented use and machine efficiency.

This section provides guidelines to assist in the design and development of codes which support both features.

Ten characteristics of a good code

The most viable and useful coding system is one which contains the greatest number of the following ten features:

Uniqueness

The code structure must ensure that only one value of the code with a single meaning may be correctly applied to a given entity or attribute, although that entity or attribute may be described by name in various ways.

Expansibility

The code structure must allow for growth of its set of entities or attributes, thus providing sufficient space for the entry of new items within each classification. The structure must also allow classifications to be expanded and others added as required. Generally, a doubling of the original set should be accommodated, with normal expansion between presently assigned positions.

Conciseness

The code should require the fewest possible number of positions to include and define each item. Brevity is advantageous for recording, communication, transmission, and storage efficiencies.

Uniform size and format

Uniform size and format is highly desirable in mechanised data processing systems. The addition of prefixes and suffixes to the root code should not be allowed, especially as it is incompatible with the uniqueness requirement. Because such prefixes and suffixes are often of variable length and do not always appear, inconsistencies and confusion result.

Simplicity

The code must be simple to apply and easily understood by each user, particularly workers with little experience.

Versatility

The code should be easy to modify to reflect necessary changes in conditions, characteristics, and relationships of the encoded entities: such changes must result in a corresponding change in the code or coding structure.

Sortability

Reports are most valuable for user efficiency when sorted and presented in a predetermined format or order. Although data must be sorted and collated, the representative code for the data does not need to be in a sortable form if it can be correlated with another code which is sortable.

Stability

Codes which do not require to be frequently updated also promote user efficiency. Individual code assignments for a given entity should be made with a minimal likelihood of change, either in the specific code or in the entire coding structure. Changes are costly, laborious, cause errors, and can also degrade the system if they are uncontrolled.

Meaningfulness

As far as possible, codes should be meaningful. Code values should reflect the characteristics of the encoded entities, such as mnemonic features, unless such a procedure results in inconsistency or inflexibility.

Operability

The code should be adequate for present and anticipated data processing both for machine and human use. Care must be taken to minimise the clerical effort and computer time required to continue operations.

It should be noted that these features may conflict. For example, if a coding structure is to have sufficient expansibility for future needs, it may have to sacrifice conciseness. Hence, all trade-offs must be appropriately considered, to enable optimum efficiency within a given structure.

Planning a coding system

When designing a new coding system, sufficient effort and time must be spent in preliminary study, definition and planning. Potential problems must be anticipated and all design alternatives thoroughly evaluated *prior* to implementation of the new system.

When properly used, significant codes can provide additional information and are easier and more reliable for human use than non-significant codes. However, caution must be exercised in the development of significant codes to ensure that the significant parts are connected to stable entities. Obviously it is necessary to avoid designing a code which can be wrecked by a minor reorganisation. Code significance is often overvalued. Researchers looking for a lower error rate amongst users have failed to find it. Significance is generally helpful only to a limited set of uses, particularly human identification and interpretation activities.

It is wise to use existing code systems where possible. Apart from the upheaval and effort involved in designing and introducing new codes, there are considerable costs involved in the creation of directories and reference files.

More than one code or representation is necessary to meet most systems requirements. A single code is the ideal objective, but it is not always the most practical solution. Multiple codes, if needed should be translatable from one code to another, ie the data items remain unchanged, only the codes are variable.

Mnemonic codes may be used to aid association and memorisation, thus increasing human processing efficiency. Mnemonic structures must be carefully chosen to ensure that flexibility is not sacrificed. Where mnemonic or otherwise meaningful codes cannot be provided for all codes in the system, preference should be given to codes having the highest use frequency.

To calculate the capacity of a given code for all situations while maintaining code uniqueness, the following formula applies (assuming 24 alpha characters and ten numeric digits are used: the letters 'I' and 'O' should be avoided wherever possible):

$C = (24^A)(10^N)$

where

C = total available code combinations possible

A = number of alpha positions in the code

N = number of numeric positions in the code (A + N, when combined, equal the total positions of the code).

NOTE: The above formula assumes that a given code position is *either* alpha *or* numeric – never both. If a given position can have *both* alpha and numeric characters, the formula becomes:

$C = (36)^{A+N}$ or $(34)^{A+N}$ when the letters 'I' and 'O' are not used.

Code length

Codes should be of minimum length to conserve space and reduce data communication time, but at the same time optimized in terms of the code capabilities and its expansibility.

A code of fixed length is more reliable and easier to use than a variable length code. Codes longer than four alphabetic or five numeric characters should be divided into smaller segments for purposes of reliable recording, eg XXX-XX-XXXX is more reliable than XXXXXXXXX. Five characters is the maximum size for a code or code *segment* which people will find easy to remember.

Code format

Code components and segments should be formatted according to user needs for information, and ease of scanning for accuracy, completeness, and compactness of the message. Message formatting should be coordinated among system users.

The choice between alphabetic and numeric codes is decided by machine capability, cost and error rate.

Machine capability is seldom a problem, except, of course, for equipment such as OCR, MICR, bar code readers with a limited or non-alphabetic capability.

The cost is generally favourable to alphabetic codes since they can be shorter (eg five alpha and eight numeric positions have about the same capacity).

The question of error rates is complex. For each letter transcribed, the probability of error is greater than for each numeric digit. However, the error rate is more sensitive to length of code than whether it is alphabetic or numeric. As alphabetic codes can be shorter for the same capacity, this reduces the overall error rate and more than compensates for the higher error expected of individual letters. Thus:

- if alphabetic codes can be used to give shorter codes, then use them;
- if the code is likely to be the same length regardless, then use numeric codes.

In cases where the code is structured with both alpha and numeric characters, similar character types should be grouped, and not dispersed throughout the code. For example, fewer errors occur in a three character code where the structure is alpha-alpha-numeric (ie, HW5) than in the sequence alpha-numeric-alpha (ie, H5W).

If a code divides an entire entity set into smaller groupings, the high-order positions should be broad, general categories, and low-order positions should be the most selective and discriminating (including any prefixes and suffixes). An example is date (YYMMDD). If a descriptive code is formulated consisting of two or more existing independent codes, the individual code segment occupying the higher-order position will be based on usage requirements and processing efficiency considerations.

Code segments should be separated by a hyphen (when displayed), or exist in complete separation (when stored and displayed) if the positions or segments are completely independent and can stand alone, (ie, no other code is required for complete meaning).

When the number of characters of a proposed code exceeds four and will be used for purposes of identification of major subjects (eg organisations, projects, materials, individuals) then consideration should be given to the addition of a check character to detect errors. Employment of a self-checking code avoids many problems of processing incorrect data.

Character content

Familiar characters should be used, and symbols (eg hyphen, period, space, asterisk, etc) avoided in code structures (except for separating code segments where a hyphen may be used). Capital letters only, should be used in data codes. Names and abbreviations may use both upper and lower case letters, and symbols. The vocabulary for a given code system should contain the fewest possible character classes. Wherever possible, the character set used for data standards should conform to the ISO 646: 7-bit Coded Character Set for Information Processing Interchange.

When it is necessary to use an alphanumeric random code structure, characters that are easily misread should be avoided, eg letter I for number 1 or symbol /; letter O for number zero; letter Z for number 2; and letters O and Q, U and V, and A and H.

Non-significant codes should avoid characters that can be confused when pronounced (acoustically homogeneous); for example, the letters B, C, D, G, P, and T or the letters M and N.

Avoid the use of vowels (A, E, I, O, and U) in alpha codes or portions of codes having three or more consecutive alpha characters to preclude inadvertant formation of recognisable words.

Assignment conventions

Quantities or numbers should not be coded since this produces additional translation and a loss of preciseness. For example, the numbers 1 to 99 could be coded A, 100–199 coded B, etc. This may be desirable for purposes of categorisation, but statistical value is lost since the actual numbers cannot be derived once they are coded. Categorisations can be performed during later phases of data processing rather than in precoding of the input data.

To maintain fixed code length and avoid the confusion of leading zeroes, codes assigned in sequence may be assigned beginning with '101', '102' or '1001', etc, rather than with '1'. Another advantage of this practice is keeping unauthorised persons from determining the quantity of data in the total entity set from knowledge of a single code (eg Product Serial Numbers). Code Numbers with lower values may be used to identify miscellaneous or special situations, if so desired, or may be left unassigned. (This procedure does reduce code set capacity, however.) All '0's' (implies nothing) or all '9's' (implies the end) should not be assigned code values. These values should be reserved for special situations or for use as processing indicators. A code category for 'Miscellaneous' or 'Other' varieties must be used with great discretion. One should not allow the placement of entities in this category which actually belong in a more specific class.

Finally, the systems analyst must, in designing the coding system, bear in mind two most important aspects: who will apply the code and in what environment, and who will interpret the code and in what environment. The function of the code, its use in searching or identification or classification, will be greatly affected by the people involved with it.

SUMMARY

This chapter presents a description of various coding systems with a brief comment on their usefulness in particular situations. The two main categories of code are significant and non-significant and within these categories there are several types, but it should be emphasised that most codes in practice will combine elements from more than one of the types discussed. The systems analyst will rarely have to design a major coding system because the costs involved in introducing such a change are too high for it to happen frequently. When a new code is required, various principles can be followed. The coding system should be designed to offer uniqueness, expansibility, conciseness, uniformity, simplicity, versatility, sortability, stability, meaningfulness, and operability. Special attention should be paid to code length, code format, character content and assignment conventions. Codes have become especially important with the advent of computer systems; data which was formerly handled in an unstructured way has to be highly structured to allow manipulation; thus, code design has a major impact on both manual and computer subsystems.

18 Designing User Procedures

INTRODUCTION

The advent of the computer to the user environment brings about greater clarity in procedures and provides the opportunity for better control. User procedures must be designed to ensure the former and take advantage of the latter. It is not enough to deliver to the user precisely what was asked for and then to assume that the procedures will cope.

In a fully clerical operation procedures can and often do develop gradually, but in a computer-based system, user procedures must be carefully designed to interface with the machine and vice-versa. In a batch processing system, the inputs must be available at the right time and in the right format for data preparation to take place. In an on-line system different constraints apply but there is still a need for greater formality in formats and procedures.

This chapter is concerned primarily with procedures within the user department which lead up to computer input and which make use of computer outputs. The systems analyst is often required to advise on staffing and resource levels within a department as a whole, and so in certain situations will be concerned with procedures which are not computer-related. The design of user procedures is closely related to forms and dialogue design as well as computer procedure and computer input/output design. This chapter concentrates on specific aspects of user procedures, in particular, work flow, office layout, staffing and equipment levels, work measurement, error handling and organisation structures.

WORK FLOW

Work flow design is concerned with identification of tasks to be carried out and determination of the most effective sequence and the most appropriate division of labour. It is quite ineffective to break a procedure into a number of simple steps and make one person responsible for each step, on the model of flow line production. Such a design is likely to be counter-productive for two reasons.

Firstly, a job which is repetitive in short cycles is boring and so people either will not stay or will become error prone. Secondly the cumulative delays between the various steps expand the throughput time out of all proportion to the time required for the individual transaction. As an example, on an incoming customer order, it may be necessary to:

- check that catalogue numbers are specified;
- query with the customer some item not fully specified;
- check the product prices specified on the order and take up any queries with the customer;
- ensure that the discount terms specified by the customer are correct;
- ensure that the delivery address is fully specified;
- transcribe to a punching document.

If each of these operations is allocated to a different person, the actual operation time may not vary greatly from that which will be taken if one person was doing the whole job, but there will be the addition of five 'out-tray' and five 'in-tray' periods. If each of these is kept as low as fifteen minutes, then an order which could have passed through the system in less than an hour will be taking between three and four hours. In a non-computer environment this could cause delays in fulfilling the order. With a computer working in batch mode the delay in input might be entirely unacceptable.

Documenting the flow

Once the workflow has been determined, it must be documented both for discussion with the user department and to form a basis for the User Manual.

If there is a simple sequence of operations with no documents entering or leaving other than at the beginning or end, and no branches in the procedure, then narrative may be a satisfactory way of recording. Long sentences and paragraphs should be avoided, and each instruction should begin on a new line. If there is any complexity, then some more formal method is essential. The most usual one is the flowchart, but decision tables or grid charts may be useful in clarifying a complex inter-related set of decisions or operations.

Flowcharts, decision tables and grid charts have been described in Chapter 7 (vol. 1) when the investigation of procedures was being discussed. The same principles apply when using these tools of documentation in the design stage. Clerical procedure flowcharts can be drawn in a variety of ways (by procedure, by location, by file) to highlight different aspects of the procedure.

Decision tables are only used in clerical procedures when a compact illustration of a complex decision is required. A document/department grid chart can be used to show the use of the same document by different departments; and a document/item grid chart which shows the occurrence of items of data on different documents will highlight duplication of work and redundant information.

All the documenting aids should be used in the design process as working documents to assist the analyst and the users to design effective procedures. They can help in the questioning of procedures under the following headings:

Necessity – what purpose does the activity serve? What would happen if it didn't exist?

Sequence – is the activity in the best sequence in relation to other activities? What would happen if it was placed at different positions in the overall flow?

Combination – could this activity be usefully combined with any other activity (thereby, cutting down waiting time and making a job more interesting)?

Simplification – is there any way in which the activity could be improved, for example by the provision of mechanical aid?

The critical examination may well involve a detailed look at the layout of the office or offices where the procedure is taking place. Where there is a lot of movement of documents, the facts recorded on the Document/Department Grid should be transferred onto a plan of the office. Where there is a regular movement of documents from one part of the office to another, and this movement is necessary because of the different specialisations of the staff concerned, there is a *prima facie* case for rearranging the office to bring those two sets of staff nearer together. This should not be undertaken lightly: probably both physical and human problems will need to be considered.

OFFICE LAYOUT

Occasionally the systems analyst will be required to design an office layout. In doing so, certain objectives should be borne in mind. Firstly, the layout should be designed to achieve minimum movement of work. It should be possible to avoid staff having to travel even short distances frequently to pass on work or to access files, etc.

Secondly, the layout should make maximum use of the available space; in certain offices this can be difficult because of the building structure. Thirdly, the design should facilitate supervision, ideally so that supervision can be less close but more easily achieved. The fourth aim should be to reduce environmental interference and to ensure that the maximum conditions of health and safety are observed. Finally, any likely expansion in size of operation must be taken into account.

Planning the layout

The plan to be used in any layout changes should show windows, doors, radiators, telephone and power points. A lot of time will be saved in drawing and redrawing if cardboard templates are made of desks, filing cabinets, etc, which can then be moved about on the plan until a satisfactory solution has been obtained.

This still leaves the human factor. People form attachments to other people, to objects, and to positions. A move may mean no longer being within chatting distance of a particular friend or group. Someone who is being moved away from the window may object to being deprived of the light or the view, while someone who is being moved nearer the window may object to the draught. This does not mean that existing positions must remain forever, but it does mean that the person initiating the change must be sensitive to possible problems.

Environmental aspects

There are several aspects of the physical environment of clerical procedures which should be taken into account when examining office accommodation. The most common ones are:

- level of lighting;
- temperature and humidity;
- noise levels;
- colour schemes;
- seating and furniture (desks, bookcases);
- machines and equipment;
- security and privacy arrangements;
- prestige and status;
- access and communication;
- electrical supplies;
- toilets and cloakrooms;
- canteen facilities;
- dirt and cleaning facilities.

Reference should be made to relevant legislation such as the UK Offices, Shops and Railway Premises Act. This includes minimum provisions for heating, ventilation, lighting, safety, fire hazard, position of toilets and number of people per square foot (or, with low ceilings, per cubic foot) of office space.

Flexible layouts

Where the layout designer has a large room available for accommodating staff, then usually an open-plan design is most appropriate. Open plan offices make use of demountable partitions instead of permanent walls to separate groups of staff. This enables greater flexibility in the use of space, eases communication, and is cheaper per person accommodated because space is not wasted by corridors or walls. It does however require efficient heating and air conditioning facilities, and also furniture designed to reduce noise levels. Also security requirements may demand special offices.

STAFFING AND RESOURCE LEVELS

Requirements for staff, terminals and other office equipment need to be determined at a fairly early stage. Staffing levels are impossible to specify until procedures have been designed in detail and tested; but physical resources, such as photocopiers, calculators or filing equipment, can usually be identified in advance.

Where equipment is dependent on document volume rather than number of staff (eg photocopiers), firm predictions can be made. With equipment which is dependent on number of staff (eg telephone extensions, typewriters), precise numbers may well not be specified until the system is actually in operation.

Human factors in user procedure design

The capability of a *computer* can be defined and procedures specified accordingly. Specifying *human* procedures is considerably more complex. One may start from an 'ideal' procedure, which assumes that people with a defined range of abilities will be operating it, but there may be people already in posts for whom employment has to be found or there may be a shortage of the defined range of abilities among job applicants. It is therefore important to know of any such constraints before starting to design new user procedures.

This does not mean that every procedure should be made so simple that a child could carry it out; performing such tasks may be boring to adults. If some or all of the existing staff will be the people concerned, it will help if the analyst who carried out the original detailed fact-finding can be closely involved with the design of the new procedure: there should be some awareness of the level of intelligence and degree of flexibility of the people concerned, and the personal relationship should instil confidence. There should be a somewhat lower degree of resistance to change than there would be to ideas advanced by an unknown person.

Ideas from the investigation stage may still have relevance and, if the people in the user department who put forward those ideas can be involved in further development, the probability of success will be greatly enhanced. Firstly, someone doing the job is likely to have a better grasp of the day-to-day problems than an outsider; secondly, people will always enter more wholeheartedly into systems that they have helped to develop than into those which are imposed from the outside.

Staffing levels

Once the user system is running, care needs to be taken that the number of clerical staff is neither too big nor too small. If it is too small, meeting the required schedules will always be a problem; if it is too big, Parkinson's Law (work expands to fill the time available) will operate. People without enough to do find other ways of occupying their time. Over-staffing can therefore, in time, become as much of a problem in meeting schedules as under-staffing. In addition it is more expensive.

Since the systems analyst will usually not be trained in work measurement, it is advisable to call on the services of O & M or Work Study specialists, who should have the necessary skills. If such skills are not available, then the analyst should reach agreement with the head of the user section concerned and with the clerks concerned regarding the number of transactions per hour.

WORK MEASUREMENT

Work measurement is the quantitative analysis of the time taken to do a particular job. It is mainly used to determine workloads for staff and to achieve some equity of distribution of work. This facilitates more precise planning and control of activities. Work measurement is also used to set up working standards which can then be used as a basis of comparison between both jobs and workers, and to highlight exceptions and unusual events.

The normal procedure adopted in work measurement is to begin by assessing the volume of work in a particular section or activity. Then the time taken to do this amount of work is measured by observation and timing for set periods, and subsequent estimation of total time required. The observed timing has to be adjusted in the light of judgement of the effort expended, the pace of work and comparison of the speed and accuracy of different members of staff. Various work measurement systems are in use by O & M specialists, including Variable Factor Programming, Group Capacity Assessment, Clerical Work Improvement Plan, and

Activity Sampling. The systems analyst is advised to call in the specialist to apply these techniques.

ERROR HANDLING

The control aspect of data processing systems is discussed in Chapters 12 and 14. The cause of errors and the steps that can be taken to reduce errors are now considered.

On the whole, errors are not the result of lack of ability or conscientiousness on the part of user staff. They are usually the result of poor system design. Most errors are caused by boredom in the tasks to be carried out (eg too much checking), excessive copying (leading to transcription errors), complex calculations (inadequately simplified) and difficulties of classification (having no clear rules by which to analyse/classify data). When errors occur frequently, they should be investigated to discover their nature and volume, their location, and their cause; and effort should be devoted to improving the system in order to reduce the errors.

The main steps which can be taken to reduce errors will involve redesign of part of the system. It may be that a form has been badly designed or a code is difficult to apply. The user manual may be unclear and the instructions inadequate; staff may have been poorly prepared and trained; supervision may not be tight enough or may be too tight; jobs may have been badly designed, not giving sufficient autonomy and responsibility to the individual; the control procedures may be lacking or poorly designed for identifying errors; or perhaps inadequate use has been made of automatic checking facilities such as check digit verifiers, and edge-punched cards.

The onus is on the systems analyst to design a system which helps the user to avoid making errors. The analyst should decide on a tolerable error level and take steps to investigate any problems when the errors exceed this level.

It cannot be emphasised too much that the performance of the users in the system will be directly improved by their involvement in its design, proper training before the new system starts, and a well-written and well-presented user manual.

ORGANISATION STRUCTURE

The design of user procedures will invariably bring about changes in the nature and scale of work in the user department; usually methods of work will be changed, sometimes social and environmental conditions will change, and occasionally the people involved will change. These changes demand that the organisation structure of the user department be reviewed to see whether it is meeting its changed objectives. It is not appropriate here to examine the various theories of organisation and leadership, but among the issues which must be considered by the systems analyst are:

- responsibilities of different members of staff;
- reporting and accountability structure;
- decision areas and levels of decision;
- span of control of managers and supervisors;
- delegation;
- relationships with line and service departments;
- information flow and communication (hierarchical and lateral);
- motivation and job satisfaction;
- formal and informal relationships;
- planning and control procedures;
- education and training arrangements.

The approach to reviewing the organisation structure must start with identification of organisational objectives and examination of management tasks in relation to the achievement of objectives. This will point to the work required and methods of organising this work including delegation of decisions to lower levels in the organisation. The analysis of tasks will point to the type of staff required and specification of their work. Finally, mechanisms should be set up for co-ordination of the tasks identified and for regular review of performance.

One of the major aspects of organisation structure which frequently confronts the system designer is whether to centralise or decentralise the tasks and staff involved in the system. Centralisation tends to be more economic (in terms of usage, staff, equipment and space) and allows more staff opportunities for specialisation, promotion and training. It also facilitates more even work loads, reducing duplication and problems of communication. On the other hand, decentralisation promotes higher morale, better overall job knowledge, more control over priorities, and less interruptions/distractions. In other words, both methods have advantages and disadvantages and the choice must depend on the particular circumstances of the system under consideration. Usually a solution which combines the good points of each is to be preferred.

Finally, it must be emphasised that the systems analyst can only ever be in an advisory role. Decisions about organisation structure will be made by the departmental manager. Often the analyst will have no more than a peripheral involvement.

SUMMARY

Before the advent of computer-based information systems, procedures tended to evolve over a period of time; now, however, because of the interface with the computer they must be specified in detail. The systems analyst is concerned with the flow of work in a department, the layout of offices, the levels of staffing and equipment, work measurement and the handling of errors. The manual subsystem must be designed to optimise these factors. In addition, the systems analyst will often be required to make recommendations to a departmental manager about appropriate organisation structures for the new system. In all these aspects of user procedures, the aim must be to increase the job satisfaction of the user department staff.

Part VI
System Implementation

Once the physical system has been designed in detail, the next stage is to turn the design into a working system and then to monitor the operation of the system to ensure that it continues to work efficiently and effectively. The implementation stage of a project is often very complex and time-consuming because many more people are involved than in the earlier stages. This part examines the issues involved.

Chapter 19 discusses the preparations which are necessary before implementation can begin; these include planning the implementation, training the user staff and testing the system. Once the preparations are complete, the changeover itself can begin and this is covered in the next chapter; the main issues are file conversion and set-up, and choice of changeover method. Chapter 21 deals with the problems of ensuring that the new system operates as required; this involves maintenance activity as a result of amendments and regular reviews of the system.

19 Preparation for Implementation

INTRODUCTION

Implementation is the stage of the project when the theoretical design is turned into a working system. At this stage the main workload, the greatest upheaval and the major impact on existing practices shifts to the user department. If the implementation stage is not carefully planned and controlled, it can cause chaos. Thus it can be considered to be the most crucial stage in achieving a successful new system and in giving the users confidence that the new system will work and be effective.

The implementation stage is a systems project in its own right. It involves careful planning, investigation of the current system and its constraints on implementation, design of methods to achieve the changeover (including writing programs to convert files), training of staff in the changeover procedures (as well as in the new system procedures), and evaluation of changeover methods.

The more complex the system being implemented, the more involved will be the systems analysis and design effort required just for implementation. Indeed in large organisations systems analysts may specialise in implementation system activities.

Tasks of implementation

The tasks involved in the normal implementation process are shown (fig. 19.1). The first task is implementation planning, ie deciding on the methods and timescale to be adopted.

Once the planning has been completed, the major effort in the computer department is to ensure that the programs in the system are working properly. At the same time the user department must concentrate on training user staff. When the staff have been trained a full system test can be carried out, involving both computer and clerical procedures. Planning,

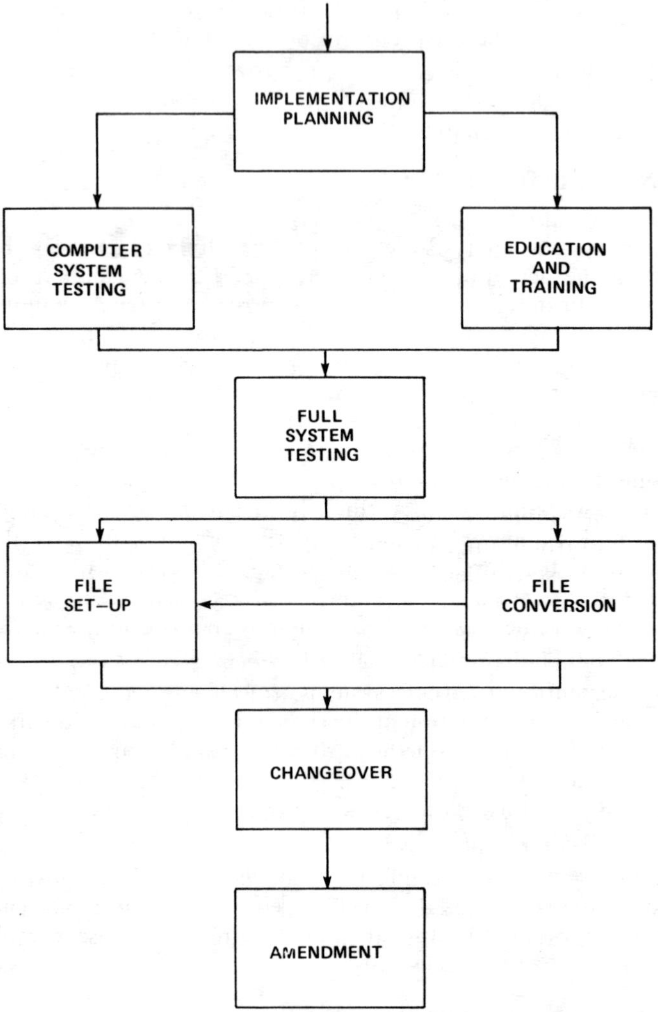

Figure 19.1 Tasks of implementation

training and testing, considered in this chapter, are the preparatory activities of implementation.

Once the preparation is complete, the implementation proper can begin. The first part of this involves the conversion of existing clerical files to computer media and the setting up of these files as they are converted on the computer. Then the actual changeover from the existing system to the new system takes place. These aspects are considered in the next chapter.

Finally, when the changeover has taken place, there will be need for amendments to correct or improve the new system, and this aspect is considered as part of Chapter 21.

PLANNING AND CONTROL

The implementation of a system involves people from different departments, and systems analysts are confronted with the practical problem of controlling the activities of people outside their own data processing department. Prior to this point in the project, systems analysts have interviewed department staff with the permission of their respective managers, but the implementation phase involves the staff of user departments carrying out specific tasks which require supervision and control to critical schedules. Some of these people may also need to be employed on the tasks of implementation in addition to their own present departmental tasks.

Who, then, should be responsible for a successful implementation? Ineffective control of implementation can result in the failure of a system which is otherwise excellent. Systems analysts do not have 'line' authority and even if they had, it would probably arouse resentment and resistance in both departmental managers and their staff at a time of critical importance. The systems analyst provides a service to management and cannot take executive responsibility without prejudicing this advisory role. Many of the reasons for not having a systems analyst in control apply equally to the manager of the data processing department. Vesting control directly in the line managers of user departments raises problems because of their lack of data processing knowledge.

A practical compromise is to have control vested in the line managers through an implementation coordinating committee with an advisory role, leaving systems analysts free to carry out their proper function. Where a project steering committee is already in existence, a subcommittee may be deputed to carry out this function.

The implementation co-ordinating committee

The use of a committee is a mixed blessing. It allows all aspects of problems to be considered and resolved by the appropriate line managers, particularly where the new system cuts across old or new departmental boundaries. However, it can also make decision-making a lengthy process. As with all committees, its success depends upon experienced chairmanship.

The composition of the committee is important. Usually the best person to chair the meetings is the line manager whose department is most affected by the system. There should be at least one representative of each department affected by the changes, and other members should be co-opted for discussions of specific topics. The data processing manager and the systems analyst or project leader responsible for the development of the system should be members. A line manager may be seconded to work directly with the systems analyst during implementation, and either is a suitable choice for the post of secretary. The frequency of the meetings will be influenced to a large extent by the system analyst's plans and schedules.

The committee itself will not normally exercise any executive authority; this belongs to the individual line managers who will however, be influenced by conclusions reached in committee. The committee will set up a realistic timescale for implementation and monitor progress. In the event of any executive problems which cannot be resolved, the committee would be expected to report, through the chairman, to a higher executive authority, such as the computer development steering committee.

The committee will be a sounding-board for ideas, complaints and problems in the user department, but it should also actively consider:

- the implications of the system environment;
- staff selection and allocation for implementation tasks;
- consultation with unions;
- resources available (staff and equipment);
- standby facilities;
- channels of communication;
- alternative methods of achieving changeover.

The appointment of an implementation coordinating committee depends on policy at individual organisations. In the absence of such a committee, the problems of responsibility discussed here are still relevant, and an awareness of these can assist systems analysts in planning their approach. The active cooperation of the respective managers or their representatives as discussed for the formal committee, preferably working as a team, is still required for successful implementation.

It is not suggested that the implementation process must wait until all other activities have been completed. Throughout all phases of the project, the systems analyst will be planning the activities of implementation.

Some of these activities must begin earlier than others because they take longer to complete. Some depend on the completion of others before they can be started. They also involve the use of resources and in some cases may stretch over a considerable length of time. Their coordination and control require careful planning and the establishment of schedules.

For successful implementation, systems analysts should formalise their requirements by preparing an implementation plan using the techniques described in the planning section of Chapter 23. In some organisations the implementation plan is a mandatory procedure laid down by management. A network diagram is useful to show the relationship of activities in their sequence in the overall plan, while bar charts can be used to schedule the lower level tasks for each activity on an individual basis.

With or without the formal establishment of responsibility for implementation, systems analysts will be responsible for scheduling the tasks to be done and monitoring progress. This will test their qualities of persuasion and tact.

EDUCATION AND TRAINING

To achieve the objectives and benefits expected from computer-based systems it is essential for the people who will be involved to be confident of their role in the new system. This involves them in understanding the overall system and its effect on the organisation, and in being able to carry out effectively their specific tasks. As systems become more complex the need for education and training is more and more important.

Staff selection

The implementation of a new system involves people; successful implementation depends upon the right people being at the right place at the right time. Planning networks and charts can show the number and type of people required at the place and the time, but successful implementation requires staff selection and training for that part of the system for which the staff will be responsible. Trained personnel will be needed as soon as the implementation activities begin, so training must begin before that stage.

Selection for training must take place at an early stage. Broad estimates of the numbers and types of people required should be submitted to management for approval. Then, as development proceeds, these estimates will be progressively reviewed and improved. The managers of the departments concerned, both user and operations, define the type and quality of people required for the implementation and operation of the new system, with advice from the project team. In large organisations, the personnel manager will naturally also become involved.

Like most other aspects of systems analysis and design, decisions on staff selection and training are not all made instantly, but are progressively refined as more information about the system requirements becomes available. Some requirements may be known early in the project, but there comes a time when all these requirements need to be firm.

For implementation to be successful, consideration of people who will be affected should be a high priority of the coordinating committee and the systems analyst. They should be told at the earliest possible time why the changes are necessary and how they will be affected. It is important that this communication takes place through the proper channels, ie their respective departmental managers, and not systems analysts or a committee.

It is usually advisable, and in some cases essential, for the personnel manager and trade union officials to be consulted about redeployment and especially about any redundancies. This needs to be done through the proper channels and as early as possible. It is a management responsibility, although the systems analyst may be involved as an adviser.

Systems analysts should ensure that staff selection and training schedules take account of any constraints. These may relate to availability of appropriate staff, lecture rooms, and equipment.

Training

Training requirements are easy to determine. They arise directly from the changes which the systems analyst is bringing about. User managers must be informed of how the whole system works, its objectives, new documentation, files and procedures. New jobs may be created, existing jobs changed or eliminated. User staff must be instructed in how to perform their new tasks. These requirements will be set out in principle in the User System Specification and in detail in the User Manual described in Chapter 22. Thus, the new system can be explained to user management and new tasks specified in job specifications, with the systems analyst on hand to answer any queries which may arise. But this can be arid material and to rely on manuals alone for training is to court disaster.

Training sessions must aim to give user staff the specific skills required in their new jobs. Thus they should contain large elements of practical activity (people learn far more by doing than by watching or listening). Ideally they should consist of short and regular sessions rather than a long, once-off period, and they should be conducted with small groups rather than the full number of staff; this facilitates learning from peers.

The training will be most successful if conducted by the supervisor with the systems analyst in attendance to sort out any queries; new methods will gain acceptance more quickly in this way. This also applies to user manuals; if the users write their own manuals they will be more intelligible and more acceptable.

Plenty of job-aids should be provided to assist in the learning process; these will include visual presentation of procedures (eg in flowchart form), notices on machines, wall-charts (eg of timescales), and use of different colours (eg on forms or switches). Above all, the training sessions should be tailored to the learning process, allowing assimilation over a period before more new information is fed in; and providing feedback on progress.

Education

Education is complementary to training. It brings life to formal training by explaining the background to the changes and the reasons for them. It helps to overcome the resentment that may be caused by the computer seeming to take away responsibility from individual departments.

Systems are now developed by specialists from outside the department. 'First-time' users in particular need to appreciate the contribution of computers, how they work, and how they can assist clerical functions and decision-making.

Large numbers of clerical staff have been content with their jobs. They understand the work and their responsibilities, and do not welcome changes associated with computers. To them, computers have a reputation for creating unemployment and taking the skill out of jobs. There is little evidence that computers cause unemployment, though they do cause redeployment.

Staff may be reassured when shown how computers can remove the drudgery from tasks and can cause jobs to be upgraded. If a system analyst's design produces less-skilled jobs for clerical staff, they should ask why these cannot be performed by the computer. Education should reassure user staff by placing their problem in perspective, by dispelling myths and mystique, and by indicating the contribution that computers can make to organisation objectives and to job satisfaction.

Education involves creating the right atmosphere and motivating user staff. Education sessions should encourage participation from all staff, with protection for individuals from group criticism. The sessions should make creative use of worries about the public image of the computer, perceptions of the systems analyst's role, rumours about previous implementations, etc; and they should be vehicles for the natural evolution of ideas about the system and its justification.

Education should start well before any development work to enable users to maintain, or to regain, the ability to participate in the development of 'their' system. A spirit of cooperation will be extremely valuable to the systems analyst, and will go a long way toward the successful implementation and operation of the new system. The analyst is continually in an educating role. In each meeting with users, they should be encouraged to build up their understanding of the new system. Nothing can help the user more in learning about the system than participation in its design.

Educational information can also make training more interesting and understandable. Instructions to perform new tasks should not be taught 'parrot fashion'; they should be explained within the context of the system, or the part appropriate to the task. This wider understanding of their tasks will assist user staff to deal correctly with the unusual events which will inevitably arise. The aim should always be to make individuals feel that they can still make an important contribution, to explain how they can participate in making system changes, and to show that the computer and computer staff do not operate in isolation, but are part of the same organisation. Computers produce information on which managers base their decisions: all staff influence the quality of this information by the quality of their own individual contributions within the computer-based systems.

Education can be planned for the whole organisation, separately from training for individual projects. Computer education should be included in the organisation's normal training programme for management and staff. This should start with the involvement of directors and senior managers and be presented by specialists from outside or within the organisation. From within, it is likely to be arranged by data processing management or training specialists who understand the requirements of top management.

Once education is established at the senior level it is likely to be more effective with lower levels of management, supervisors, and clerical and shop-floor staff. Data processing staff also need to be educated and trained, and should participate in such a company-wide programme.

The analyst's responsibilities

The systems analyst has a clear responsibility to press for full and adequate education and training for user staff and to ensure that provision is made for this in project budgets. This does not mean that the analyst should necessarily organise and run the training, but be advisory to user managers, providing appropriate technical expertise when called upon.

If the analyst does organise the training sessions special attention must be paid to the user's perception of the change: courses should reflect the user's view of the world rather than the analyst's view.

Finally an effort should be made to measure the success of training objectively and to learn from failures. If a member of staff is unable to cope with the new system, it will usually be the system, or the job design, or the training programme which is at fault, and rarely the individual.

SYSTEM TESTING

System testing is the stage of implementation which is aimed at ensuring that the system works accurately and efficiently before live operation commences. In principle, system-proving is an on-going activity throughout the project. The logical design and the physical design should be thoroughly and continually examined on paper to ensure that they will work when implemented. Thus the system test in implementation should be a confirmation that all is correct and an opportunity to show the users that the system works.

When the programmers have tested each program individually, using test data designed by themselves, and have verified that these programs link together in the way specified in the computer run chart to produce the output specified in the program suite specification, the complete system and its environment must be tested to the satisfaction of the systems analyst and the user.

The systems analyst will provide the test data, specially designed to show that the system will operate successfully in all its aspects and produce expected results under expected conditions. The tests should take place as far as possible in the actual operating environment, and they should test people and equipment as well as programs. Where this is not possible, the system should be tested in a simulated operational environment to prove that the computer and clerical procedures are understood and produce the required results. Sometimes it is convenient to use live data from a previous system cycle, but this presupposes that the new files have been set up and can be used. Preparation of test data and the checking of results should be carried out in conjunction with the appropriate user and operations departments.

System test data

There should be careful planning of how the system will be proved and the test data designed. The systems analyst should be quite clear about the test objectives. System test data can rarely be comprehensive enough to test the system fully; some aspects of the system will have to be tested during live operation.

Usually the test data produced is sufficient to test most of the system, whereas to test it fully would be increasingly time-consuming and difficult. This is sometimes referred to as the 'Pareto' or 20/80 effect, where 20% effort can produce 80% of the results.

Systems analysts should deliberately plan the extent to which the system should be tested. This will depend on the purpose and sensitivity of a system. Many statistical results, eg sales analyses, are not so sensitive and do not demand the accuracy needed for, say, a labour force piecework bonus, customer accounting, or product-making formulae. These factors should already have been identified and taken into account in planning the detailed design. To take an extreme case, where the system is a 'life or death' type, the extra effort and time required for the production of the most comprehensive system test data would be essential.

The system test data, and the results of processing it, should be maintained as a permanent manual throughout the operational life of the system, for audit purposes or to test any subsequent major amendments.

A suggested checklist of the contents of this manual appears in Chapter 22. Figure 19.2 shows an example of how decision tables can greatly assist in the formulation of system test data.

Test Requirements Decision Table PLO 1 6.1 TRANS-1 1

TABLE 1

C = 4
A = 6
R = 7

	1	2	3	4	5	6	7
Order no. valid	Y	Y	Y	Y	N	N	N
Transaction group valid	Y	Y	Y	N	Y	N	N
Order category valid	Y	Y	N	-	-	Y	N
Group/category comparison valid	Y	N	-	-	-	-	-
No error – go to table 2	X	-	-	-	-	-	-
Invalid category within group - EXIT	-	X	-	-	-	-	-
Order category error - EXIT	-	-	X	-	-	-	-
Transaction group error - EXIT	-	-	-	X	-	-	-
Order no. error - EXIT	-	-	-	-	X	X	-
Transaction type error - EXIT	-	-	-	-	-	-	X
	TD/TA1/1	TD/TA1/2	TD/TA1/3	TD/TA1/4	TD/TA1/5	TD/TA1/6	TD/TA1/7

Figure 19.2 Test data (decision table form)

SUMMARY

Implementation is the key stage in achieving a successful new system because, usually, it involves a lot of upheaval in the user department. It must therefore be carefully planned and controlled. Normally, this involves setting up a co-ordinating committee which will act as a sounding board for ideas, complaints and problems. Apart from planning, the two other major tasks of preparing for implementation are education and training of users and testing of the system. Education of users should really have taken place much earlier in the project when they were being involved in the investigation and design work; at the implementation stage the emphasis must be on training in new skills to give staff confidence that they can cope with the new system. Once staff have been trained, the system can be tested; it is important for the whole system and its environment to be tested and not just the computer programs. Once the co-ordinating committee is satisfied with the training and the testing, changeover can begin.

20 Changeover

INTRODUCTION

Once all the preparatory work of implementation has taken place – the system has been tested and the staff trained – the changeover from the old to the new system can begin. This involves three separate, though closely linked, activities. First of all, the old (existing) system files need to be converted to the format and content required by the new system. Then the converted files need to be set up on the computer. Finally the old procedures have to be replaced by the new ones.

FILE CONVERSION

File conversion is a vital activity which is sometimes underestimated. It involves the conversion of the old file data into the form required by the new system, and is usually a very expensive stage in the whole project. It is rare that the old files can be converted as they stand because of format changes and the need to collect and assemble data from a variety of sources. But the major problem of file conversion is that it usually has to be achieved within very tight time constraints.

Although it is usually regarded as a part of changeover, in fact file conversion is often a complete and separate system task in itself, involving fact-finding, analysis, data capture, the design of clerical methods and computer processes, form design, and the production of special training courses.

Setting up new master files for large systems can involve the transfer of tens or hundreds of thousands of records, which may be beyond the data handling capacity of an organisation and must be subcontracted elsewhere. If the task is to be carried out within the organisation, it is essential that adquate advance notice be given. Either of these methods requires detailed planning and monitoring of progress to clearly defined schedules.

The conversion of data on clerical documents invariably demands punching and verifying. It may also be necessary for the data to be first assembled and then written onto specially designed conversion punching documents.

A common problem is that the data contained on the source documents needs to be edited, either because it is irrelevant to the new computer files, or because it needs to be expressed in a different format. This requires trained clerical staff, who often have to be released from the user departments to perform these duties or to supervise temporarily recruited people, if the data is of a special technical nature.

The conversion of data held on punched or magnetic media raises different problems. Again, the data is likely to need editing and restructuring, for which a special editing computer program will be needed. If the new system has been designed to operate on a new computer, there may also be a problem of incompatibility between the coded data on the old and new magnetic media. This should have been anticipated and arrangements made with the computer supplier, but it must be taken into account when planning and scheduling file conversion.

Live files

Whatever methods of conversion are used, it must be remembered that usually they will involve the conversion of live files (eg stock files, ledger files). This poses major organisational and scheduling difficulties, since incoming data (eg stock issues, ledger payments) is continually being used to update the files. The difficulties will vary according to:

- the number of records and amount of data to be handled;
- the frequency with which the data changes.

These two factors are interrelated. Large and dynamic files increase the difficulty of capturing changing data. Small files may be released and the clerical updating process halted during the conversion process. Even in this relatively simple case, the timing is critical. Ideally, the actual conversion and setting up new files should be done at the last possible moment, so that the time and the amount of data to be changed between the start of conversion and the changeover to full computer operation is kept to a minimum.

With large files this may not be possible, the files being available only in a piecemeal fashion. In this case, a method must be devised whereby control can be maintained over a carefully planned release of the data, either in parts or sections.

A record must be kept of data received during conversion and the changeover, but which has not yet been used to update the files. This record is then used to ensure that all these data changes are eventually made to the new system files, and also to the old clerical files according to the changeover method. This calls for critical control, both for scheduling and the control of accuracy.

The conversion of large files may be assisted by first separating and converting the static data part of each record on the files, converting the dynamic contents as late as possible, and then merging the two parts of each record to make up complete records and complete new files.

Added problems occur when files have to be brought together from several locations for a centralised conversion procedure. Not only is the control problem magnified, but often the file records from the different locations are in different formats. In this case, the old files will require further processing before the conversion activity can take place.

FILE SET-UP

'File set-up' is the process of creating the new computer file from the converted, computer-acceptable data. Sometimes the file amendment programs, written for the new system can be simply modified to accept the converted data; but this approach assumes that the converted data is in exactly the same form as a new record in the new system. Usually, special programs are required to carry out some 'once only' conversion processes.

The major problems associated with file set-up are the accuracy of the conversion and the error detection and correction procedures. It is essential that, at the end of file set-up, the users are satisfied with the new files. This is a testing time for the new system, because if the new files contain many errors, the users, rightly or wrongly, will claim that the old files were always absolutely accurate. If users lack confidence in the new files, they will soon express a desire to be back on the old system.

Accuracy

It is vital that the data content of master files at changeover is accurate. Otherwise, not only will errors have to be identified and corrected during the operation of the new system, but it may also inconvenience users and cause them to suspect the design of the new system. Incorrect data may arise:

- as errors in the original source documents;
- during clerical transcription;
- during punching;
- from a conversion program.

Other errors can arise if records are lost during their removal for conversion or if some data in certain records are found to be non-existent (eg carried in the heads or in the personal records of the clerical staff).

Error detection and correction

Steps can be taken to detect errors either by computer or by clerical checking of printed files. For large files, a special computer program should be written to check the accuracy of the data. However, there are limits to the extent to which values can be checked in this way, and wherever possible, clerical checks of the printed new files should be made against either the old files or the transcription documents. The control of data, discussed above, will need to be extended to cover the correction of data, using both computer and clerical controls of total value, numeric count, hash totals, and check digits (preferably in identifiable small batches).

This is not an easy operation, but the results make it worthwhile. The production of new files of data which are demonstrably more accurate than the original files would make a good start for the new system, particularly in terms of user confidence.

CHANGEOVER

The changeover from the old to the new system may take place when:

- the system has been proved to the satisfaction of the systems analyst and the other implementation activities have been completed;
- user managers are satisfied with the results of the system tests, staff training and reference manuals;
- the operations manager is satisfied with the performance of equipment, operations staff and the timetable;
- the target date for changeover is due.

The changeover may be achieved in a number of ways. The most common methods are direct changeover, parallel running, pilot running, and staged changeover. They are illustrated diagrammatically (fig. 20.1). Occasionally a combination of these methods will be used.

Direct changeover

As this term implies, this method is the complete replacement of the old system by the new, in one move. It is a bold move, which should be undertaken only when everyone concerned has confidence in the new system. This presupposes a well organised and supervised implementation.

When a direct changeover is planned, system tests and training should be comprehensive, and the changeover itself planned in detail. This method is potentially the least expensive but the most risky. Where possible, a time should be chosen when the work of the organisation is slack. In any event, the busiest time should be avoided.

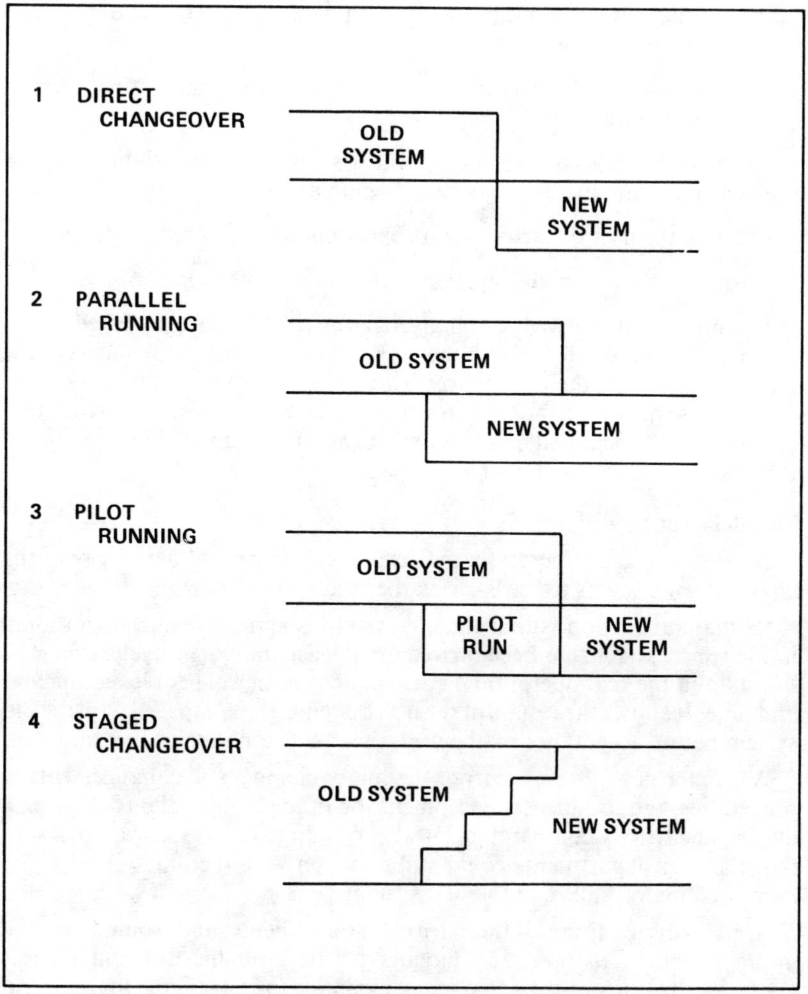

Figure 20.1 Diagrammatic illustration of changeover methods

When planning a direct changeover, it may be decided that the irrevocable disbandment of the old system before the new system has successfully completed its first full cycle is too risky. This would be a reasonable consideration when using new equipment, possibly untried under operational conditions. For security reasons, the old system may be held in abeyance, including people and equipment. In the event of a major failure of the new system the organisation would revert to the old system. Direct changeover is likely to be used:

- for a system which is new to the organisation, as distinct from a replacement system;
- where the new system incorporates such major innovations that a sensible comparison is not possible in parallel operation;
- where the user departments are experienced in computer systems;
- for a relatively small changeover or a short timescale.

It is important that written changeover instructions are prepared by the systems analyst to inform both the user and operations departments of the methods by which the changeover is to be effected and of the work to be done. These may take the same form as the User and Operations Manuals including any documents, procedures, programs and schedules specific to the changeover. (See Chapter 22.)

Parallel running

Parallel running, or operation, means processing current data by both the old and new systems to cross-check the results.

Its main attraction is that the old system is kept alive and operational until the new system has been proved for at least one system cycle, using full live data in the real operational environment of place, people, equipment and time. It allows the results of the new system to be compared with the old system before acceptance by the user, thereby promoting user confidence.

Where the new system incorporates major changes and enhancements in procedures, the use of new equipment, or in the results, the two systems may not be strictly comparable. Parallel running in this case, would ensure that the user departments could still carry on with the old system in the event of a major failure of the new system.

Its main disadvantage is the extra cost, the difficulty and (sometimes) the impracticability, of user staff having to carry out the different clerical operations for *two* systems (old and new) in the time available for *one*; and then cross-checking the results, including error-handling, at a time when user staff are fully occupied with the new procedures. The use of additional temporary staff may be neither possible nor desirable. Because of these difficulties, there is a danger of neither system being properly conducted, which defeats the objective of this method of changeover.

It is not suggested that this method does not produce an effective changeover, but that the implications should first be evaluated and the operation carefully planned in detail. As it places a heavy load on the users concerned, this should be fully explained to them beforehand.

The plan should be specific about the extent of cross-checking involved, eg whether on a sample basis, the criteria on which acceptance is to be judged, and the limit on the number of cycles of operating the two systems. Errors found by the comparison are as often due to defects in the old as in the new system, but it may prove difficult to convince users of this.

Where this method of changeover has been agreed, it must not be used as an excuse to curtail proper system testing. Parallel operation does not allow much time for learning or testing activities which should have been done earlier. It can, however, in a well organised implementation, build up users' confidence as they see that the system works and that they can do their job within it.

Pilot running

Pilot running is similar in concept to parallel running. Data from one or more previous periods for the whole or part of the system is run on the new system *after* results have been obtained from the old system, and the new results are compared with the old. It is not as disruptive as parallel operation, since timing is less critical. However, users still have to cope with the clerical procedures for both the old and new systems. It does not simulate day-to-day timing and scheduling problems, and data capture and error handling are not realistic. This method is more like an extended system test, but it may be considered a more practical form of changeover for organisational reasons.

Staged changeover

A staged changeover involves a series of limited-size direct changeovers, the new system being introduced piece-by-piece. A complete part, or logical section, is committed to the new system while the remaining parts or sections are processed by the old system. Only when the selected part is operating satisfactorily is the remainder transferred.

This method reduces the risks inherent in a direct changeover of the whole system and enables the analyst and users to learn from mistakes made as the changeover progresses. Its disadvantages are that it creates problems of controlling the selected parts of the old and new systems and it tends to prolong the implementation period. It therefore needs good co-ordination and is most appropriate when the changeover has to take place at a number of different locations, each location being subjected to a direct changeover, or for very large files.

Controls

Whichever method is adopted for the changeover from an old to a new method, a high priority must be given to establishing controls, by value or quantity, in order to maintain the quantitative integrity of the system.

Users should keep overall control records incorporating both computer and clerical control figures to prove that the changeover has not corrupted this integrity. This is particularly important for financial systems, the controls for which, including audit trials, should have been planned after prior consultation with the accountants and auditors.

These controls are vital, and the tasks required to establish them can be difficult and time consuming. For large and complex systems, the most difficult task is often proving and establishing accurate controls for the old system before the changeover. This difficulty may be an existing problem, and one of the major reasons for the new system: the terms of reference should specify the extent of the systems analyst's involvement with the controls of the old system. When the accuracy of the old controls have been established, they must then be organised in a way that is consistent with the arrangement of the controls of the new system.

This aspect of controls at changeover cannot be overemphasised since not all existing systems or their control methods are in a good state of order.

HAND-OVER

Once the system has been working for an agreed period of time, the systems analyst will wish to withdraw. Prolonged involvement of the systems analyst with a working system should be avoided. This does not mean that the user department will cease to receive support. The system becomes the responsibility of a maintenance group within the computer department instead of the development staff. This hand-over point should be established as part of the implementation plan.

The users must be satisfied that the system works properly and meets all their requirements by the time hand-over takes place. It is essential, therefore, that the hand-over takes place formally, with a clear understanding on all sides that the systems analyst's involvement has come to an end.

SUMMARY

Changeover is the stage of moving over from the old manual system to the new computer-based system. In order that this can be done, the clerical files have to be converted to computer format and media and then input to the computer to form the new computer files. This is a very difficult task to achieve quickly and accurately, especially if the files to be converted are in regular use; but the accuracy of the conversion is of utmost importance both to user confidence in the system and to effective operation. When the files have been set up on the computer, the changeover proper can take place. There are several possible methods of doing this, eg direct changeover, parallel running, pilot running and staged changeover. The implementation co-ordinating committee must choose the most appropriate method and then oversee its execution. When the users are satisfied with the results, the new system can be handed over to the users and the systems analyst can withdraw from the scene. The new system is now in operation.

21 Maintenance and Review

INTRODUCTION

Provision must be made for environmental changes which may affect either the computer or other parts of computer-based systems: such activity is normally called 'maintenance'. It includes both the improvement of system functions and the correction of faults which arise during the operation of a system.

Maintenance activity may require the continuing involvement of a large proportion of computer department resources. For computer installations which have already developed the basic applications for the organisation, the main task may be to adapt existing systems in a changing environment. Perhaps a better term to describe this activity is 'system evolution'. All systems are dynamic and subject to constantly changing requirements. Effort must be devoted to adapting them, and design should be flexibly specified so that such changes can be easily implemented.

Most changes arise in two ways:

- as part of the normal running of the system when errors are found, users ask for improvements, or external requirements change;
- as a result of a specific investigation and review of the system's performance.

This chapter looks first at amendment procedures and then at formal system reviews.

AMENDMENT PROCEDURES

Systems should not be changed casually following informal requests. Changes in one area may affect others. For example, computer programs derived from man-months or man-years of careful design effort can be easily corrupted by unauthorised tinkering. Unless each amendment is properly documented in the appropriate files (eg system files, program files, reference manuals), subsequent amendments may be made which do not take into account the effects of previous amendments. Without effective control procedures, the system and its documentation will soon deteriorate.

To avoid unauthorised amendments, all requests for changes should be channelled to a person nominated by management. The nominated person should have sufficient knowledge of the organisation's computer-based systems to be able to judge the relevance of each proposed change. Whoever is responsible for initiating the amendment process must determine which system is involved. It is possible that one amendment may affect more than one system, eg where the output of one system becomes an input to another, or where a file is used in more than one system. The specific files documents and procedures affected should be identified and cross-referenced to procedure and data specifications within the master system file. Some order of priority should be given to each amendment from a scale agreed by management, eg *emergency, immediate, by a specific date, discretionary.*

Major amendments should be treated as small-scale development projects, requiring resources, terms of reference, planning, scheduling and controlling. The tasks can be specified, allocated to individuals, scheduled and controlled to target dates.

When operational runs fail, system errors may be corrected by standby emergency staff. The amendment documentation can be initiated at the time of the failure, showing the priority as *'emergency'*, defining the cause of failure and specific action taken. The documentation should then pass through the normal channel to be formally verified and authorised, and to allow amendment to system documentation.

All amendments should be completed according to their priority and scheduled target dates, especially where a discretionary priority rating exists.

Procedure requirements

In defining local procedures, certain general objects may be identified:

- any member of staff should have the means available for raising an amendment;
- any amendment must receive a suitable level of authorisation;
- each amendment must be scrutinised by someone not involved with the original proposal;

- a procedure should be designed to minimise delay in implementing the amendment;
- anyone holding documentation files must always be kept up-to-date with relevant amendments;
- the procedure must be capable of handling several concurrent amendments with different timescales;
- the procedure must not violate existing channels of communication.

Because of the various possible amendments, several procedures may be necessary within a single installation. Procedure selection and definition will depend on local priorities which determine the importance of the amendment. The following factors should be considered:

- effect of the amendment on the objectives of the system;
- effect on existing development schedules if the amendment were authorised;
- present state of the system (development, projected or operational);
- cost of implementing the amendment (it may also be relevant to consider the penalty cost of not doing so);
- number of staff to be involved in implementing the amendment;
- timescale for the change (when it is required and how long the work will take);
- source of the amendment;
- complexity and size of the changes required;
- effect of the amendment on other systems;
- effect of the amendment on users or other parts of the organisation;
- effect of the amendment on data security.

Amendment documentation

Most data processing departments have forms for authorising and controlling amendments. Examples of three such forms are:

- Amendment Notification (fig. 21.1);
- Amendment Log (fig. 21.2);
- Amendment List (fig. 21.3).

Each amendment must be accurately defined and justified. The *Amendment Notification* is designed for this purpose; it provides, inde-

Amendment Notification
NCC

Title	System	Program	Component	Amendment no
Agent code on LATES-1 list	SOP			004

Originator	Verified
Export Manager	B.J. Robson
Author	
M.L. Palmer - Systems Dept.	
Authorised	**Priority**
J.K. Johnston	Await Audit One.

Distribution
System & Prog. Masters
Audit 1 projected development file

Purpose of the amendment
Addition of Agent code to Export Orders late list will save approx. ½ hr. clerical search time each computer run.

Est. requirement: 2 man-days, 50 units m/c time.

Amendment specification

Print select program: LATES-1 set up must include extraction of Agent code from A.7/CUSORA/2 field ref 31.

Output record: A.6/RLATES-1/1 must include Agent code (amended layout attached)

Print-3: Heading adjustment for LATES-1 (A.3/LATES-1/1)

S11

Figure 21.1 Amendment notification

Amendment Log NCC

Title	System	Program	Component	Amendment no.
Agent code on LATES-1 list	SOP			004

Activity	Estimated completion	Responsibility	Actual completion	Comments
Specify	12/6/77	MLP	12/6/77	
Appraise & Design	1/7/77	BJR	1/7/77	No spec. modifications
Code & Compile	7/7/77	BJR	9/7/77	High priority of other wk.
Test	15/7/77	BJR/MLP	16/7/77	
Complete & Handover	17/7/77	MLP	17/7/77	Approved x Export Mgr

Document	Amendment Details
P120/3/SEL07/2	Select Agent Code for printing
SOP/4.6/RLATES-1/1	Add Agent Code
P140/2/HEAD-2	" " "
SOP/4.3/LATES-1	" " "

S10

Figure 21.2 Amendment log

Amendment List NCC		System	Document		Sheet
		SOP	9.2		1

Amendment No.	Title	Date of issue	Letter of issue	Documents affected
004	Agent code on LATES.1 list	17/7/77	B	4.3/LATES-1

S12

Figure 21.3 Amendment list

pendently of the author, for verification and authorisation of the amended specification.

The *Amendment Log*, prepared intially with the Amendment Notification, records the estimated and actual completion of each of the activities defined by the author.

The *Amendment List* provides a permanent record of all the amendments made to any of the documents within a documentation file: it is a useful reference to keep at the end of each file. The serial number is the amendment number entered on the corresponding Amendment Notification. The original issue of any document can be identified by the letter 'A'; each time a document is replaced by an amendment the issue letter is changed sequentially, in the series B to Z, AA to ZZ.

Maintenance group

Responsibility for the maintenance of a particular system must be allocated before any requirement for changes arises. It is unwise for this responsibility to rest with the original designer after 'changeover'. If the number of amendments justifies a separate maintenance team, then this team usually reports either to the operations manager or to the programming manager. The maintenance team should be allowed to influence the original design of systems/programs insofar as it will affect their 'maintainability'.

SYSTEMS AUDIT

The systems audit is an investigation to review the performance of an operational system: to compare actual with planned performance; to verify that the stated objectives of the system are still valid in the present environment; and to evaluate the achievement of these objectives.

This investigation and evaluation may be carried out: by a systems analyst, preferably one who was not responsible for the original design; by representatives of users, computer operations, or internal auditors; or by a team composed of these representatives. A knowledge of systems design is essential for analysis of findings.

The initial review should take place when the system has had time to settle down, when any additional assistance by systems analysts and temporary staff is no longer required, when both equipment and people are operating satisfactorily, and before any major changes are made to the original design specification. This is unlikely to be less than three months after changeover. The initial systems audit provides the opportunity to check whether the objectives and benefits forecast in the feasibility study have been achieved. Subsequent audits, carried out as part of regular reviews of systems (perhaps annually) will be concerned with the continued achievement of benefits, any deviations from the master system specification, and opportunities for improvement.

The detailed tasks to be carried for this investigation are based on a checklist of the contents required for the Systems Audit Report (see Chapter 22). They are summarised under two main headings:

- system performance;
- cost/benefit.

It must be emphasised that the over-riding reason for an audit is to verify that the stated objectives of the system are being achieved, or that they are still valid in the present environment. These objectives (for which management authorised the use of resources in the first place) must be established before attempting to evaluate the system performance. The objectives and cost/benefits will be found in the management report of the feasibility study, and in subsequent reports. The expected performance of the system, in broad terms, should also be found there, but the System Specification should be referred to for details.

System performance

The investigation should start by making contact with the manager of the user departments, not only to deal with the normal formalities but, in particular, to establish:

- whether or not the manager is satisfied with the performance of the system, and if not, what are the reasons;
- the use being made of the output reports; whether they are accurate; if they are timely; whether they contain insufficient or unwanted information;
- the operational aspects: whether the procedures are causing problems, and if any changes have been made;
- effectiveness: if there are many error reports; whether there are inaccuracies not being reported; turnround and response times; the level of equipment utilization, reliability and service;
- changes in volumes of data, information, paper handling and their effect on the system;
- amendment requests: whether they have been implemented correctly; whether there are any pending.

The above facts should also be established in more detail from other levels of user staff employing the procedures.

When the personal interviews have been concluded, the auditor should quantify actual performance to establish any deviations from the planned performance, together with explanations: this is the main objective of the audit. Deviations, which may be classed as avoidable, can arise from incorrect estimates, eg:

- clerical and computer procedure timings;
- data volumes and growth rates;
- staffing levels;
- error rates.

Costs/benefits

Here the actual costs and benefits are compared with those planned, showing any deviation from expectations. The causes of any deviation of costs or benefits from those planned should be established and stated. These may arise from:

- unplanned pay increases;
- extra staffing, retention of initial temporary staff, inaccurate estimates;
- changed methods of computer charging;
- inaccurate estimates of data volume and timing;
- authorised, or unauthorised changes to procedures and documents.

Deviations may be either advantageous or disadvantageous, and all details should be reported. An increased cost may, of course, produce a better service, perhaps giving higher value.

Other types of deviation are usually environmental, arising from operational, trading or statutory changes, eg:

- pay methods, accounting policy, new equipment and techniques;
- production and selling methods;
- product and market standardisation or diversification;
- organisational expansion or contraction;
- government statutory returns, new taxes.

Where changes already have been made to the system, these should be summarised together with the causes. A check should be made that these have been officially approved and have followed the correct amending procedures, particularly that the changes have been recorded on the master system file and in the other appropriate reference manuals.

The performance statistics should first show the comparison of actual with planned, and only then should the effect on these of any amendments and improvements be shown. These comparison records can be fed back to the planning and estimating section, to system analysts and programmers, to improve future forecasting methods.

Management requires to know of any benefits which have not been realised, and the causes. To enable a realistic comparison to be made, the criteria used initially as a basis for estimating the financial benefits should be employed. Where these are found to be ineffective in any way, reasons should be given, and alternative criteria then applied to both the estimates and the actual evaluation. This is particularly applicable to the criteria used to evaluate intangible benefits. Any additional benefits arising which were not expected should be noted; and any financial gains arising from changes made since the changeover. There should also be comment on the likelihood, of attaining any long-term benefits.

Quality assurance

The level of control within the system deserves special attention. It is essential that adequate control procedures are built into the system as it is designed. These should be checked to ensure that they are working effectively, one being maintained, and that the system is secure. The following checklist gives a summary of the questions which should be asked.

Control environment

- is there clear segregation of control responsibilities?
- have all mandatory controls been specified?
- what standby procedures are there, and what is the cost of having to use them?
- can user involvement be adequately demonstrated?
- how will the user monitor system operation?
- are there procedures of authorisation to check the quality of data?
- are documentation standards maintained?

Source data collection

- have batch sizes been defined and maintained for maximum control?
- have batch control records been defined and maintained?
- are batch controls established as soon as possible?

- is the data collection environment suitable?
- are data collection and verification procedures clearly specified?
- are the following procedures defined and maintained correctly:
  - registration of receipts?
  - verification of receipts?
  - recording of work distribution?
  - data conversion controls?
  - error procedures?
  - procedures for special equipment?

Validation

- is all input verified to the required standard before processing?
- are all fields validated for range, format and size?
- are check-digits used where appropriate?
- are input records verified for completeness, content and field sequence?
- has sufficient use been made of possible field comparison tests: consistency, credibility, cross-checking, acceptability?

Error control

- are control reports adequate both for errors and successful runs?
- are any errors processed with acceptable data, and if so are suitable safeguards included?
- do error-override facilities exist, and are there special controls to prevent their misuse?
- has the most appropriate method been used for clearing a validation run?

Computer procedure and file controls

- does the program design include contingencies for overflow, timing and frequency variations, environment changes, and variations in machine conditions?
- are file controls adequate in the form of labels or special control records?

- are control totals kept for all significant fields, plus record counts and hash totals?
- could separate control files be usefully kept?
- is there a full reconciliation system linking input data to all output?
- is there appropriate provisioning of an audit trail?

Output procedures

- do output programs validate new fields created for output purposes?
- are key fields rechecked for credibility?
- are fields and reports edited according to standard?
- are full control reports provided?
- are reconciliations reported whether successful or not, with supporting control totals?
- are program performance figures reported?
- are output control reports produced?
- is all output approved and monitored by an output control section?
- is there an output control register?
- are control totals verified by output control section?
- are special procedures defined for confidential data?
- does the user receive a control report showing costs, volumes, and error-rates?
- are periodic in-depth checks performed on output data?

Use of terminals

- does terminal dialogue have built-in redundancy for error detection?
- is clerical data input minimised by the use of machine-generated data, or avoidance of fixed data keying?
- is input distinguished from output?
- are there adequate message controls (serial numbers, logging, hard copy, audit trail)?
- is the terminal design best for this application in terms of keyboard, screen format, security, etc?

- what data protection facilities are provided for data transmission?
- are error correction facilities provided through retransmission, reconstruction or using the principle of no acknowledgement?

Fallback and recovery

- are fallback clerical input procedures defined?
- is there a specified procedure for re-establishing controls?
- are all messages logged on receipt, and are these logs retrievable for recovery purposes?

External requirements

- have all appropriate external authorities been consulted?
- has the auditor approved the system controls?

Recommendations

Ways to improve system performance should be given: either to meet or exceed expectations. If additional work is needed, then the terms of reference should be formulated in detail.

SUMMARY

Even when the new system has gone live there may be need for some system design activity. This will stem from changes that are necessitated by the dynamic nature of the system and its environment. Changes may be required to correct faults or to bring about improvements, and may arise as part of the normal running of the system or as a result of a review of the system performance. Changes (or amendments, as they are called) must be carefully controlled by appropriate procedures and documentation and perhaps handled by a specialist group of maintenance staff. The review of system performance (system audit) will usually take place when the system has settled down and will be concerned primarily with looking for improvements in the performance of the system and ensuring that it is achieving the forecasted benefits. The system audit will also examine the level of control in the system. It may be that the result of one of the annual audits will be to recommend a complete redesign of the system and so the cycle of development will start again.

Part VII Project Documentation and Management

This final part of the book looks back at the various project activities which have been described in the previous chapters on the investigation, analysis, design and implementation of systems. It is clear from the earlier material that system development is quite a complex activity, especially if large or integrated systems are involved; it, therefore, needs careful control and this is the concern of this part.

Chapter 22 provides a detailed summary of the various reports that the systems analyst may be involved in producing at particular stages of a project; these range from the Study Proposal, which initiates the system development, to the System Audit Report, which looks at whether the developed system is effective. The final chapter of the book, chapter 23, then discusses the importance of project planning and control in achieving successful system development. A number of planning and control aids are discussed.

22 Project Reports

INTRODUCTION

At various stages in the development of a system, the systems analyst has to communicate ideas about the system. Proposals must be submitted to management for approval; users need to know what the system will do and what is required of them in terms of clerical procedures; operations staff need instructions for running the computer system. This chapter identifies the types of report required and gives checklists of their contents. Checklists provide guidance and a high standard of presentation.

The various system documents included in the reports described in this chapter will be derived from the *system documentation files:*

- Old System File, containing all documents defining the system as it exists at the time of the investigation;
- New System File, containing all documents defining the system as proposed, agreed or implemented at any particular point in time;
- System History File, containing all ideas and proposals which have been considered and rejected, all superseded documents, and all amendment notifications.

The most important of these is the New System File which, when completed, contains the specification of the new system. It is built up during the stages of the project, and designated as one of three editions:

- *Development* (material which is not yet operational);
- *Operational* (documents describing an operational system);
- *Projected* (documents describing a system which is fully developed and tested but not yet operational).

Within the New System File are all the procedure specifications (including flowcharts, decision tables, network charts, etc); data specifications (including clerical documents, file and record specifications, print layouts etc); relationship charts (data structures, program structures, grid charts etc); and test documentation for the system. The file is structured in accordance with the filing references shown (fig. 7.1, vol. 1).

From the documents included in these files, the following project reports are produced:

- *Study Proposal*
- *System Proposal*
- *User System Specification*
- *Program Suite Specification*
- *User Manual*
- *Operations Manual*
- *Test Data File*
- *Changeover Instructions*
- *System Audit Report.*

STUDY PROPOSAL

The Study Proposal is the first project report: it sets out a case to management for the use of resources to undertake a study of a system (usually a feasibility study). It is concerned only with the work needed to carry out preliminary investigations, which lead to a formal System Proposal or result in a report recommending no further action. The Study Proposal ideally should be produced by, or on behalf of, a user department, normally before project work commences. The contents of the Study Proposal should be the minimum necessary to enable management to make an informed decision whether to authorise the work; it should therefore highlight the justification, costs and timescales without prejudging the outcome of the study. The following is a structured checklist of contents.

Title page:

report title and reference;

author and department;

month and year of publication;

space for authorisation;

distribution list.

Summary:

brief description of the nature of the proposal;
origination of the proposal;
costs of the study;
anticipated completion date for the study;
(the summary should not extend beyond one typewritten page.).

Proposed terms of reference:

description of the problem(s) or other requirement;
the purpose and scope of the study;
constraints on the study in terms of cost, timescale, resources;
constraints to be placed on the outcome of the study;
reporting mechanisms: method, timing, method, including progress control.

Resource requirements:

manpower required for the study;
departments directly or indirectly involved in the study;
additional resources anticipated, eg consultancy, computer time;
support services required, eg accommodation, typing.

Timetable for the study

Organisation and membership of the study team

SYSTEM PROPOSAL

Once the study has been authorised by management, the investigation of the existing system or problem area can commence in accordance with the terms of reference. The result of this work will be another report to management, called a System Proposal. This sets out the case for making any change to an existing system. This may involve:

- deciding initially whether to use a computer;
- approving extensions to existing applications or equipment;
- sanctioning work leading to new applications or equipment.

A System Proposal will not be updated. However, during analysis and design work there may be a series of such documents each more detailed and specific than the previous one, stating progress to date and describing the next stage of work requiring authorisation. For example, depending on the method of project organisation, System Proposals could emanate first from an initial study (ie, feasibility study), and then one or two from detailed design work, each in greater depth than the previous one. The first Systems Proposal in any series will have resulted from work authorised through a Study Proposal. The final report in any series may, if necessary, recommend that no further work be authorised. The contents of the Systems Proposal must be the minimum necessary to enable management to make an informed decision. It should highlight the effects and implications of the proposals, anticipated costs, savings and other benefits. Details, especially technical details, should appear in the appendices. A checklist of contents might be:

Title page:
- report title and reference;
- author and department;
- month and year of publication;
- space for authorisation stamp/signature;
- distribution list.

Contents list:
- main and sub-headings with section/sheet numbers.

Summary:
- objectives and proposals showing, in succinct form, the identified needs of the user and how they will be met; indicating if the needs can be met using existing facilities or the extent to which new development and/or equipment is required;
- costs of development, implementation, operation;
- benefits;
- (The summary should not extend beyond one typewritten page.)

Recommendations:
- a statement of the management decisions required for immediate and future action assuming acceptance of the proposal;
- draft terms of reference for future work.

Introduction and scope of study:
- background to the study;
- references to previous reports;
- terms of reference of study:

- scope and objectives of the study and constraints (other systems, both existing and proposed);
- the objectives to be met by the proposal and constraints on these objectives;
- security and audit requirements;
- time and cost targets;
- modifications made during the preceding study.

Existing system:

- relevant information on the organisation and its development;
- outline and evaluation of existing system (dataflow and volumes etc);
- existing and anticipated problem areas.

System requirements:

- design requirements and constraints of a new system:
  interfaces with other systems (existing, proposed);
- rules governing operations:
  accuracy;
  quality;
  schedules;
  cost;
- evaluation criteria for the system when implemented;
- future projections, extensibility, fluctuations.

Proposed system:

- outline of system;
- alternatives considered and rejected with reasons;
- implications of the new system of interest to management:
  - details of necessary re-organisation (accommodation, staff deployment);
  - hardware and software;
  - support services;
  - training;
  - operating schedule;
  - security and/or audit;
  - insurance;
  - union interests.

Development and implementation plans:

- outline showing main features of organisation proposed, major control points, manpower and external requirements;
- broad schedule.

Costs:

- expenditure to date;
- estimate of costs to continue/complete development;
- implementation/installation costs;
- operating costs, compared with existing costs;
- (costs should be broken down into: manpower, equipment, software, support services, consumables, etc.)

Benefits, quantified where possible:

- savings on current costs;
- better resource utilisation;
- information quality;
- improved control.

Appendices:

- existing system supporting information;
- proposed system supporting information;
- description and evaluation of hardware, software, communications, environmental, power and back-up requirements;
- detailed implementation plans, including additional services required, file conversion, training, integration, changeover, controls, target dates;
- development project organisation;
- glossary of technical terms.

USER SYSTEM SPECIFICATION

Once management has authorised the start of design and development work for a new or amended system it is necessary to provide user mangement and staff with information about the main features of the system and how it will affect them. This information is contained in the User System Specification, and should form the basis for any final agreement with those who are expected to use the new system. Agreement should be reached before detailed design work is completed.

The specification should not require to be updated once the contents are agreed. Subsequently, however, it may be possible to produce the User

Manual from the User System Specification. The specification also plays an important part in user education and may be used for staff familiarisation before the User Manual becomes available. It should therefore be presented as a training aid, and not as a reference manual.

The contents checklist provided below contains cross-references to working documents. However, working documents should only be used when it is known that these will be acceptable and understandable to the user. It may often be preferable to present data or procedures in plain-language terms.

Introduction:

brief introduction to the system;

reasons for introducing changes;

statement of problems;

objectives of the new system and expected benefits.

Procedure summary:

one page summary showing major changes and explaining principles of the new system;

system outline;

system flowchart.

Procedure specifications:

description of the clerical and interactive procedures within the system;

brief description, using non-technical terminology, of the objectives of the new/revised computer procedures.

Data:

samples of mock-ups of input and output documents and displays;

specification of new clerical files;

summary description of computer-based data files, showing which files they replace.

Supporting information:

organisation chart showing lines of responsibility in terms of the new system;

document/department grid.

Changeover:

plans for change to a new system;

timescales, critical activities, and work loads.

Operations:

anticipated schedules;

deadlines and critical points.

It should not be forgotten that users will ultimately have full responsibility for running the system. Computer processing, carried out in the computer operations department, is done on behalf of users. The user departments will pay for development and operation of the system, either directly, or indirectly through the management accounting system.

Such a specification is often the users' last opportunity to request changes where the design fails to meet their requirements. Once they have accepted the specification, it is often declared to be 'frozen', ie, no amendments will be accepted during the remainder of the project.

PROGRAM SUITE SPECIFICATION

Once the final system proposal has been accepted by management the systems analyst has the task of specifying the computer functions of the system to the programmers. The Program Suite Specification must provide sufficient information to the program development team about the computer functions of the system to enable programs to be developed and tested to a defined level. The content will, therefore, vary between different installations and between different teams. Much will depend upon the point at which program development is distinguishable from system development. Other factors are organisational structure, experience of the people involved and design novelty.

At one extreme the information that has to be communicated may be little more than a statement of the objectives to be met by programming activity, together with a broad description of input, output and maintained files. At the other extreme the information may be so detailed and complete that coding can commence following receipt and examination of the specification. Where programming is a function of a multi-discipline 'project team', and there is no readily distinguished transfer of responsibility, there may be no need for a formal Program Suite Specification.

The specification will be produced to form the basis of *sub-contracting* responsibility for the programming work. The handover of the specification represents an identifiable control point in a development project and usually involves locally-defined procedures such as for authorisation, specification appraisal and acceptance, and handling of queries. Local standards must also provide for specification maintenance.

The content of the Program Suite Specification will be a subset of the documents maintained in the New System File, the subset being determined in the light of local requirements. For ease of subsequent referencing and maintenance the specification should be organised in the same sequence as the System File from which it is derived.

Local standards must specify the content of the specification. The table in figure 22.1 may be used for guidance in formulating the local standard: this recommends which parts of the System File may be abstracted to the Program Suite Specification. (Note: the table in figure 7.1 (vol. 1) should be consulted for more complete definitions of System File content.)

USER MANUAL

The purpose of the User Manual is to instruct the user departments in the clerical operations required for the successful operation of a system and to inform them of the facilities available from the computer programs, constraints on operations and actions to be taken in the event of failure or errors.

The manual is a permanent form of communication (as distinct from the user system specification and changeover instructions which may be discarded when they have served their purpose) and must be available to users before the new system goes live.

It must always be accessible for reference purposes throughout the operational life of the system and should therefore reflect the current system and be updated as and when changes are made which affect user procedures.

The manual should only contain copies of documents from the system file which are of direct relevance to the user and which are easily understood. Where necessary it should be structured so that it can be easily segmented by department, section, or even by individual function. Its structure should also reflect the needs of a reference manual; it will only be used to resolve questions of procedure as they arise from time to time. Whilst it is primarily a reference document, it may also be used as a basis for training purposes.

It should be particularly explicit in those areas where user staff may have little or no previous experience, ie, at the boundary between clerical and computer operations, preparation of input data to the computer, the handling of input and output errors, any unusual or infrequent events which might be expected to arise, and for the general control of the system.

Users who have not had previous experience of computer-based systems are not usually aware of the importance of formal and strict adherence to instructions, and the care needed to avoid errors in computer processing. This should be taken into account.

The master copy of the User Manual should be kept in the systems department files. Further copies or sections will be kept in user depart-

System file contents		Recommendation
Doc.ref.	Name/definition	
1	Background	B
	Communications	
2.1	Discussions, meetings	C
2.2	Correspondence	C
2.3	Associated documents	A
	Processes	
3.1	Overview	D
3.2	User-clerical procedures	A
3.3	Operations procedures	C
3.4	Computer process: organisation	D
3.5	Computer process: details	E
	Data	
4.1	Clerical data	A
4.2	Source data files	D
4.3	Output data files	D
4.4	Stored data files/logical database	D
4.5	Source data records	E
4.6	Output data records	E
4.7	Stored data records/item groupings	E
4.8	(Not used in System File)	–
	Support	
5.1	Analyses and interactions	C
5.2	Data item usage	D
5.3	Hardware, software, etc.	D
	Tests	
6.1	Specification of test requirements	F
6.2	Test plans	F
6.3	Test operations	F
6.4	Test logs	–
7	Costs	C
8	Performance	C
9	Documentation control	G

The recommendations are coded as follows:

A – Not usually required.

B – Probably required, but in a different form from that held in the System File. Rewriting should take into account the extent to which programmers need the information and the need for brevity.

C – Some extracts may be needed, but only of that information necessary for programming work to be accomplished.

D – Required: any changes or expansions are unlikely to be a programming responsibility

E – Required: changes or expansions might result from programming.

F – The information may be required in the form of objectives, rather than in the form held in the System File.

G – The Program Suite Specification may have its own documentation controls (eg for copying). Amendments to the specification will be subject to local standards.

Figure 22.1 Recommendations for contents of Program Suite Specification

ments. The responsibility for maintaining all copies must be clearly allocated. A checklist of contents might be:

Title page:

- title, author and author's department;
- month and year of publication;
- name, department and telephone number of contact(s) in the event of problems during the operation of the system and for general enquiries concerning use of the system.

Contents list:

- main and sub-headings with section/sheet numbers.

System summary:

- as brief as possible, and explained non-technically. It should not extend beyond one typewritten page.

Clerical and terminal input procedures:

- System Flowchart and description of the whole system, including options, part-runs, etc;
- Procedure Flowchart and procedure description for each department involved or, for smaller systems, for each function involved;
- batching, controls, error detection and correction;
- timetable for any time-critical activities.

Computer input documentation:

- completed example of each document/display facing a page of description and supported, as necessary by a Clerical Document Specification or Display Specification;
- conversion tables, codes;
- handling of incorrect/incomplete documents;
- error correction.

Computer output documentation:

- sample of each output and explanation of contents;
- distribution of output;
- description of possible error reports;
- handling of errors.

Non-computer documentation:

- completed example of each document facing a page of description and supported by a Clerical Document Specification;

- handling of faulty documents;
- error correction.

Glossary of terms:

- explanation of any technical terms which the user may be required to understand. This may include the program names.

Amendment list.

OPERATIONS MANUAL

An operations manual is used as a permanent reference document to inform the computer operations department of the system to be implemented, the work to be done in its routine operation, and any special features.

The manual is the formal communication of system details to the operations department, but is not the only communication needed. The Operations Manager and senior operations staff play an important part in system development, as a sounding board and source of technical expertise, prior to the detailed system design and production of material for the operations manual. Such material will be produced progressively as the system is developed. It is essential that provisional details be supplied to the operations department as soon as they are available to give opportunity for preparation of preliminary schedules and forward loading plans and for training and familiarisation.

The contents should be clear and factual. As it may be necessary for the manual to be partitioned to the requirements of operations sections (eg, data capture, data control, job control, operations), its structure should be determined in consultation with the operations manager. It should be designed to enable problems of operation to be solved without continual reference to programmers or systems analysts.

The master copy of the manual should be kept in the systems department's file, with other copies held in the operations area. The contents might include:

Title page:

- title;
- author and department;
- month and year of initial publication;
- implementation date of version to which manual refers.

Contents list:

- main and sub-headings, with section/sheet numbers.

Application details:
- brief description of the application including options, alternatives and exceptions to the main pattern.

Place of system:
- brief note showing the inter-relationship (if any) of this system with others.

Summary of operations:
- System Outline;
- System Flowchart
- Computer Run Chart;
- File/program grid.

Timetable:
- frequency of runs (all types);
- relationship to calendar periods/weeks;
- min/max/average duration;
- priorities;
- commencement date (actual/expected).

Computer requirements:
- store/partition size;
- input/output and backing;
- peripherals;
- job streaming (for multiprogramming);
- priority ranking (where used).

Input data:
- source of input documents;
- samples of each type of document;
- min/max/average quantities;
- punching and verification instructions and layout;
- controls, eg batching;
- arrival schedule, estimated punching times;
- Procedure Flowchart;
- who is responsible for queries/checking;
- destination of input documents when punched.

Files:
- purpose of each file;
- medium requirements min/max;
- number of reels, packs, etc;
- identifiers;
- labels to be provided before/during run cycling;
- purge dating;
- retention;
- references from/to other systems;
- security;
- physical position in library.

Output data:
- samples of each type of output media;
- off-line requirements;
- Document specification(s);
- other requirements, eg retention, collating, enveloping;
- Procedure Flowchart.

Programs:
- identifiers of programs involved, including packages and software;
- authors;
- dates of current versions;
- who is responsible for faults.

Operating procedures (normal):
- sequence of events through each run;
- controls, actions required of operator;
- full list of monitor reports, replies, with samples;
- peripheral loading and unloading.

Operating procedures (abnormal):
- full list of failure reports and actions to be taken;
- actions to be taken for other failures;
- contacts in event of failures.

Operating procedures (restart):

- actions to be taken to restart runs, after a failure, with particular reference to any other system which may need re-running prior to restart.

Amendment list.

TEST DATA FILE

Once the programs have been written and the users are familiar with the new procedures, the system can be tested to ensure that it will operate successfully under all the expected conditions and will produce the expected results.

The Test Data File is a document in its own right which is produced primarily for the use of the systems analyst, the users and the auditors. It must be kept up-to-date throughout the operational life of the system, and used everytime an amendment is made. The file should specify all the aspects of the system, data and procedures which are to be tested. It should demonstrate the expected test results and cross-reference each test to the item of test data which initiates the test; listings of test data together with specimen source documents should be kept. The test data should test the complete system to the satisfaction of the systems analyst and the users; this will include clerical and computer procedures, and inter-relationships with other systems. Thus the preparation of test data and checking of results will be carried out in conjunction with user departments.

A checklist of contents might be:

Title page:

- title and reference;
- author and department; date.

Contents:

- main and sub-headings with chapter/sheet numbers.

Testing philosophy:

- describe the approach to testing of the system;
- identify any distinct stages or timing considerations;
- if the testing involves the creation of complex conditions, explain how these may be set up;
- emphasize any aspects of the test which are likely to be overlooked, or areas which may prove troublesome or critical.

File creation:

- test data for creation programs will be in the same form as the data for setting up the live master files. Files will be checked for validity of format and accuracy of data. Check record controls and file security.

Program suite input:

- data conversion;
- data transmission/Remote Job Entry;
- data control;
- computer validation;
- error routines and correction procedures.

Program suite output:

- user acceptance of program output;
- form design;
- data control procedures;
- data transmission.

Input/output handling:

- each procedure will be checked for accuracy and understanding, ambiguity, timing, and staff confidence;
- completion of input documents;
- maintenance of clerical files;
- checking of documents;
- delivery of output and inspection;
- distribution of output;
- actioning and turn-round of documents;
- error procedures;
- contact with computer operations department.

Test data:

- test data listings and data preparation documents.

CHANGEOVER INSTRUCTIONS

Changeover instructions are necessary whenever systems are radically altered (minor alterations can be handled by the amendment procedures described in the last chapter). They are concerned solely with the work needed to build up to, and accomplish the changeover from the old to the new system; the instructions are no longer required once the changeover has been effected and should not form part of the User or Operations

Manuals. When a number of sections are affected by a changeover, it is recommended that each section receives only those instructions which are of particular relevance, rather than a comprehensive set for all sections.

Where the changeover is to be accomplished in phases it will be generally necessary to modify the procedures, etc in line with these phases. For example, the way in which certain part of file take-on are handled might vary according to how far the changeover has proceeded. In such cases it will be necessary to indicate clearly which parts of the changeover instructions refer to which phases. Alternatively, it may be less confusing to revise and re-issue the instructions as the changeover proceeds.

There are two main sets of changeover instructions, those for user departments and those for computer operations.

User Departments Changeover Instructions

These instruct the user department staff in the methods whereby changeover is to be effected. A checklist of contents might be:

Title page:
- title, author and author's department;
- month and year of publication;
- name, department and telephone number of contact(s) in the event of problems during the period of changeover and/or for general enquiries concerning the changeover itself.

Contents list:
- main and sub-headings with section/sheet numbers.

Introduction:
- brief introduction to the new system;
- reasons for introducing changes;
- outline of the method of changeover, including main schedule dates;
- brief description of ancillary services available for the changeover.

Clerical and terminal input procedures*
- procedure flowchart and procedure description for each department involved in the changeover, or, for smaller systems, for each function involved;
- controls, error detection and correction;
- start and finish dates for each stage or phase in the changeover.

Computer input documentation*

- completed example of each document/display facing a page of description and supported, as necessary, by a Clerical Document Specification or Display Specification;
- conversion tables, codes;
- handling of incorrect/incomplete documents;
- error correction.

Computer output documentation*:

- sample of each output and explanation of content;
- distribution;
- description of possible error reports;
- handling of errors.

Non-computer documentation*:

- completed example of each document facing a page of description and supported by a Clerical Document Specification;
- handling of faulty documents;
- error correction.

Glossary of terms:

- explanation of any technical terms which the user may be required to understand.

Amendment list.

* To be provided for those procedures and data concerned with effecting the changeover from the old system to the new and also for those affected by the changeover itself.

Operations Department Changeover Instructions

These inform the operations department of the conversion to be done and the work to be performed by the department. A checklist of contents might be:

Title page:

- title/author, author's department;
- month and year of publication;
- start and finish dates of changeover period.

Contents list:

- main and sub-headings with section/sheet numbers.

Introduction:

- for a new application, brief description; for a replacement, reference to the old system and a brief description of the changes to be made;
- outline of the methods of changeover and of integration with other systems.

Timetable:

- schedule of the file creation/conversion runs showing expected durations and volumes;
- schedule of any other special runs required during the changeover period;
- fallback and recovery requirements in the event of a halt being called to the changeover.

Computer requirements Input data Files Output data Programs Operating procedures	:—	the contents of these sections of the changeover instructions will be as for the Operations Manual but will describe the procedures, etc necessary to accomplish the changeover, and also any non-normal activities on other systems and on the system being replaced, during the period of changeover.

Amendment list.

SYSTEM AUDIT REPORT

When the system is operational, one other document that the systems analyst may be involved in producing is the System Audit Report. Its purpose is to report on the performance of a system (in particular to compare actual with planned performance), to verify that the stated objectives are still valid in the present environment, and to evaluate the achievement of these objectives. A checklist of contents might be:

Title page:

- title, reference;
- author(s) and department(s);
- month and year of publication;
- distribution list.

Contents list:

- main and sub-headings with section/sheet numbers.

Summary:

- brief re-appraisal of the objectives of the system;
- reference to original proposals;
- brief statement of conclusions, indicating any aspects of the system which are unsatisfactory and stating what objectives have been achieved;
- brief statement of any differences of opinion between users, designers and operations.

Recommendations (if necessary):

- proposed changes to the system or its environment and justification for the proposals;
- effects of proposals on user and operations departments;
- recommended short-term management decisions, assuming acceptance of the proposed changes;
- draft terms of reference for further work.

System performance:

- summary of all available performance statistics and comparison with estimates:
  - computer time charged/resources used related to data volumes and transactions;
  - growth rate of files and transactions;
  - manpower requirements for clerical systems;
  - turn-round times for user departments and operations;
  - efficiency of security procedures and quality control checks;
  - error rates for clerical operations and data conversion/entry;
  - delays attributable to operational problems (eg schedule clashes, hardware and software failures, program deficiencies and operating errors);
  - suitability of rerun and restart procedures, back-up and standby arrangements;
- the effect of changes in the environment on the performance of the system:
  - summary of program amendments and the causes;

  - relevance to the system of any new techniques or technological advances;
  - changes in company policy or other external influences which affect the performance of the system;
- the attitude to the system of the user at all levels from management to operative;
- reactions from customers or other external bodies;
- attitude of the computer staff;
- comparison of the use being made of computer output with the potential usefulness;
- any unplanned uses for output, or any redundancies;
- verification that superseded clerical systems have been discontinued;
- effect on related systems which have been influenced by the system under review;
- outstanding problems arising from this appraisal of performance of the system, including a statement of the degree of adherence to standards and relevance of instructions and procedures specified in User and Operations Manuals.

Cost/benefit review:

- present system operation cost;
- the acknowledged benefits both to the company as a whole, and to individual user departments;
- unplanned developments or activities which have provided additional benefit;
- any excessive costs with possible justification;
- explanation of benefits expected but not achieved;
- a subjective, independent assessment of the expected intangible benefits;
- forecast of any long-term benefits which could yet be realised.

SUMMARY

This chapter brings together suggested contents for all the major reports that the systems analyst is likely to have to produce. These are: the Study Proposal, which initiates a system investigation; the System Proposal, which can be produced in several versions at various stages in the analysis and design process depending on the depth required; the User System Specification, which is the definition of the logical system for user evaluation; the Program Suite Specification, which specifies the computer data and procedures; the User Manual which specifies the manual subsystem; the Operations Manual which specifies the details of both manual and computer subsystems to the computer operations department; the Test Data File which specifies the testing of the system; the Changeover Instructions which provide information to users and computer operations about changeover procedures; and the System Audit Report which reports on the review of the system.

23 Project Management

INTRODUCTION

This chapter is concerned with activities ranging from the management of systems development projects within the overall framework of an organisation's management control system to the self-management of a systems analyst's own project or tasks. Control procedures for project development are established in many organisations to a varying extent and under various titles (such as *Project Control* or *DP Standards*). This chapter shows how a systems analyst may plan and subsequently control those activities and resources (people, equipment, money) required for the successful development of computer-based systems.

Increasingly computer projects are expected to produce a return on money invested. This situation demands accurate estimates of development costs, system running costs, and savings arising from system operation. It is important that management have confidence in estimates.

There have been failures to meet set targets of time and cost, which have to some extent fostered the search for a formula or method which would eliminate uncertainties. There is no such formula. The external trappings of a control system at management level may conceal a lack of control at the most junior staff levels.

However, it is wrong to assume that, because the project environment is fundamentally uncertain, project control is not worth attempting. Uncertainty is relative and of many forms, and much can be done to improve control by identifying and evaluating its many aspects. The effects of uncertainty in computer projects can be minimised. Good planning and control create a climate for success; people need to be both well organised and well motivated to give good results.

THE NEED FOR PLANNING

A data processing project is usually high risk, involving considerable expenditure. Project resources (ie systems analysts, programmers, computers, etc) are expensive, and the benefits arising from a new system may take time to materialise. If a project is effectively planned it is possible to evaluate the risks and uncertainties, and to reduce them where possible. Planning is the basis for control: without plans, effective control is impossible. It is important to recognise quickly deviations from plans, and to take steps to minimise their effect.

Uncertainty in data processing plans

The best-laid plans can fail. There are many factors which can introduce uncertainty into planning, some external and some internal to the organisation. External events include:

- general economic and trade fluctuations;
- business mergers;
- government and statutory changes.

Internal factors include:

- management's anticipation and reaction to business change and opportunities;
- diversification of products and markets.

Changes on the manufacturing shop-floor may occur at any time, eg during the development of a system. As one illustration, the method of paying shop-floor operators may change from piece-work to measured day-work during the development of a payroll system.

Changes in company policy and re-ordering of priorities during the lifetime of a project can have wide-ranging implications for project planning. The effect of these changes can be reduced by both formal and informal means. Formally, the setting up of a Computer Steering Committee composed of management at a senior level ensures their involvement in overseeing computer projects. This should ensure that project teams are aware of proposed changes in sufficient time to take effective collective action. This can also be achieved by the appointment of a line manager as project leader.

Often a company is not prepared to introduce this kind of organisational innovation. In such circumstances systems analysts must find out, by means of their own informal contacts, about proposed changes in company policy.

Changes in specification during the life of a project are clearly an irritation; often they necessitate plans being changed and program amendments. Such changes often follow incomplete specifications being supplied by systems analysts to programmers; or failure on the part of user departments to state their requirements completely and accurately. The incorporation of user department staff into project teams will go some way to ensuring that user requirements are reflected in the system specification.

Documentation standards can help the systems analyst to produce specifications which are complete, so reducing the problem arising from poor specification. Even so, it is doubtful if specification changes will be eliminated completely. They can be reduced by having a formal means for requesting amendments and some mechanism for sifting the essential from the non-essential.

Long-term plans

Ideally, projects should not be planned and developed in a corporate vacuum, nor merely by duplicating experiences in other environments. They should be planned as tactical goals and constituent parts of long-term strategic plans to achieve corporate objectives.

Most organisations expect computers to improve their financial results. However, methods of achieving this aim are specific and peculiar to each industry and organisation, and they change with changing circumstances. They must therefore be determined by top management. Data processing staff can advise and apply their technical skills to realise objectives, provided that the objectives are clearly stated.

In the absence of specified corporate objectives, computers can be useful in many applications. However, there can be no assurance that they will be fully exploiting business opportunities or solving problems effectively.

At their most effective, long-term data processing plans must be developed as part of an overall strategy which corporate management accepts as a means to achieve specified objectives.

Thus as changes of policy occur, modifications made to data processing plans may be minor or involve a radical re-shaping of future plans. It is important that systems analysts and management recognise the heavy commitment in computerisation plans and management decisions formulating (and arising from) them. Once an organisation has embarked upon a particular course, and acquired an increasing dependence upon a computer-based system, it may be difficult and costly to change direction.

In the UK, most organisations studied have established a computer Steering Committee, composed of management representatives from the main areas of the organisation. The aim is to monitor and review computerisation plans and their implementation. Such plans are concerned with:

- a framework for the selection of applications, their boundaries and interfaces with one another;
- a set of objectives for management to establish a common purpose;
- overall guidelines and constraints for data processing management;
- guidance at the technical level on equipment, programming languages, system design philosophy, etc;
- a basis for data processing department budgeting.

Reporting to the steering committee will be several project teams which will carry out feasibility studies, and detailed design and implementation activities. A project team will normally consist of line department and computer department staff. Occasionally projects are controlled by a project committee, chaired by the line manager of the department most affected by the changes. The project team would then report to the project committee; this sort of structure enables line management to have a direct influence on the development work.

PROJECT PLANNING

Various factors discourage planning: it costs money; it uses the time of the most highly skilled people; the time spent in producing the plan can delay the start of the project; a plan can cause inflexibility of approach and estimating is difficult. These objections, however, do not detract from the importance of planning.

A computer project usually originates in one of two ways. As more and more companies are realising the wisdom of having longer-term plans for their data processing activities, individual projects may originate as part of these overall plans. In other organisations, projects may still originate as requests from individual line managers who feel that a computer-based system may help them.

Data processing management is responsible for the selection of projects according to agreed priorities. Each project leader is individually responsible for a given project. The purpose of planning projects is to:

- order their subsequent development to a predetermined pattern which makes the most effective and economical use of resources;
- provide a yardstick against which to control project development and resource utilisation;
- establish meaningful terms of reference which define the objectives and constraints of the project, its boundaries and relationships with other projects and systems, both working and proposed;

- minimise the risks of omissions and ambiguities in development, and their costly consequences;
- specify what is to be done, how, when and by whom, and how much it will cost.

Answers to these questions are usually sought by engaging in an initial study or series of studies.

The initial study must be brief, relative to the size and complexity of the area under study, to avoid the wastage of resources that might occur if the results of a more detailed study proved to be unacceptable to management. It is important because it selects the system area for future development, defines boundaries, problem areas, constraints, and recommends the terms of reference for a more detailed study. Because this study requires a considerable amount of judgement at a time of least information, it should be carried out by senior staff experienced in both data processing and a knowledge of the organisation. For these reasons, it is sometimes assigned to outside consultants.

The end product of the study will be a proposal to management which answers the questions posed above, states the case for a proposed system and includes a development plan for the project. Depending on the size, scope and extent of the innovation, this study may be sufficient to enable management to make a decision to proceed or not to proceed with the project. Alternatively, the initial study may identify particular problem areas which require further studies; for example, appraisal of applications packages, data capture equipment.

Initial studies can be regarded, in a sense, as ancillary to the development of the project. This is because, until such a study has been undertaken, management generally has insufficient data to decide whether or not to proceed. The initial study is often separated from the development of the project: in many organisations which normally charge users directly with project development costs, the cost of the initial study is not charged directly but is borne by the data processing department as an overhead.

Project phases

One of the ways of dealing with the uncertainty at the start of a project is to divide the project into more easily manageable parts or phases.

A project can be divided into progressive sets of activities which represent stages in the production of the new system:

- establishing terms of reference;
- fact-finding;
- recording existing system;

- analysis of the facts leading to a specification of user requirements;
- evaluation of various alternative design solutions and an outline design of proposed system;
- preparation of a proposal to management;
- detailed design of computer and clerical procedures leading to the provision of a system specification;
- preparation of a proposal to management;
- programming and program testing;
- preparation of user and operations manuals;
- user training;
- system testing;
- system implementation;
- system maintenance.

Each of these sets of activities is time-consuming; some are sequential; some take place in parallel with others. There may be interaction with other activities (such as other projects, delivery of hardware, holidays, training).

It may be possible to split some projects into parts which are either logically self-contained or special problem areas; these can be divided into some or all of the above phases.

The activity stages can be used to identify the main phases in the life of a project, for example:

- planning: initial study (proposal to management);
- development: detailed study (proposal to management); detailed design (proposal to management); programming; implementation;
- system audit: post implementation evaluation.

These phases can be clearly defined in planning, and identified during development. They are major milestones and as such are useful targets and review points. Consequently, the first three phases are each shown to be concluded with a proposal to management; this sets out a case to management for approval of progress made, and for resources to complete the remainder of the project.

In some cases, management may only authorise the resources necessary to complete the next phase; further resources for each successive stage may only be authorised after acceptance of the proposals at stipulated review stages.

When plans have been made at the beginning for the complete development of the project, management can compare actual progress and resource utilisation at each review point. They can authorise with confidence further development against this planned profile.

Estimating

Estimating times for projects is still the most difficult planning task to formalise. For example, this is the case in construction, engineering and other development projects.

There is at present no generally applicable approach. Systems analysts have to use whatever experience is available to them, by performance records. In programming projects more use is made of formulae.

The difficulty of estimating increases the need to analyse project development into more understandable and manageable phases. There is also the requirement to identify the lower-level tasks, and the activities required for their completion. Realistic estimates are important and are more likely to be achieved if based on an installation's own performance rather than on unsubstantiated abstracts from literature or other installation environments.

Resource scheduling

The allocation of resources involves matching the ability and experience of available project staff with the estimates of resource requirements. Resource time estimates have to be set or modified according to the experience of the person who will perform the activity.

The production of schedules involves determining which activities can be performed in parallel, which in sequence, and the subsequent calculation of elapsed time for the project. The difference between resource time and elapsed time must be clearly understood.

For the systems analyst, resource time is usually productive machine-time or productive man-time, and is used to describe the total resource commitment to specific tasks or even to the whole project. This may be calculated as man-hours, man-days, man-years, machine-minutes, or machine-hours, and is the time multiplied by the number of resources employed on that activity. This contrasts with elapsed time which describes the passage of time, ie calendar time, between any two points (for example, between intermediate phases, tasks or activities); it is the time estimated or taken to achieve an objective, including any time when no work is done.

Resource-time and elapsed-time for a specific project rarely have a direct relationship, unless development is completely sequential with no concurrent activities or interruptions; this is rare. Elapsed or calendar time is important for fixing target dates.

Effective resource time is that which effectively contributes to the progress of a project. It is important to recognise that sometimes a proportion of booked resource-time does not in fact do this. Much resource-time can be occupied by interruptions, discussion, assistance on non-project activities of administration, maintenance, or other project queries.

In devising schedules it is possible to produce estimates for planned, productive and unproductive times; unplanned, unproductive time is best allowed for by a percentage figure applied to planned time. It is important that this percentage figure results from a system which records unplanned, unproductive time.

Forward resource loading

The availability and allocation of resources may change from day to day during the development of projects. As a data processing department expands and more projects are authorised, the more necessary it becomes to monitor resource utilisation. In many organisations this has resulted in setting up a forward loading resource register which is: a record of the current allocation of resources to projects; and a forecast of allocation requirements for future development on a period basis. The responsibility for maintaining this register may be assigned to a systems analyst. The register should, of course, show any planned non-project activities, such as training and holidays.

PLANNING AND CONTROL AIDS

The techniques examined here do not provide fool-proof methods of planning and controlling projects; such techniques do not exist. The relevant experience of the planner is probably the most important factor in determining the accuracy of plan and estimates. But the techniques can be especially useful as means of formally recording development experience and making it available for future projects. The main aids are:

- Gantt charts;
- networks;
- network – bar charts;
- check lists.

Gantt (bar) charts

Bar charts are the simplest means of depicting plans. Each bar on the chart is proportional to the time the activity it represents will take. Columns carry a time-scale along the horizontal side; the elements or activities planned are listed on the vertical side. An open box or line is drawn on the chart to represent the planned duration of each activity, the thickness of the box is constant but its length along the chart indicates duration, since it extends between the starting and finishing times. In addition to the open box for planned time a thick line may be drawn to represent actual time taken (fig. 23.5).

Bar charts can be used to record the planned activities for all project development staff; these activities include holidays and training, in addition to project work. A loading chart is thus produced for each individual.

Since the bar chart is a scale diagram it requires amendment when estimates of activity time are revised. This results in the redrawing of bar charts. This can be a time-consuming operation. It is often more satisfactory to amend the bar chart, until it will stand no more amendments and is then redrawn. This problem can be overcome by the use of purpose-built planning boards.

However, the charts do have one significant weakness for planning: they are unable to display clearly and unambiguously the interrelationships of the various activities of a plan; that is, when the start of one task depends on the completion of another.

Networks

A network is the best method of enabling a project to be viewed as a whole, revealing the logical relationships and inter-dependencies between component activities (fig. 23.1).

Briefly, each job or activity is identified, coded by a number, letter or a mixture of both. Then a plan of the project is drawn in the form of a network or arrow diagram of these activities. This shows only the necessary logical sequence of the tasks without any resource considerations. The estimates of each activity are then related to the network to produce the most effective sequence of work which meets the target date for completion, or the shortest elapsed time.

One advantage of drawing a network is that it forces the planner to examine the project thoroughly, to define it in detail and break it down into discrete activities. The planner is compelled to think through all the various interrelationships among the separate activities and is obliged to estimate time limits for each one. This alone would justify the effort of making the network.

The information derived enables the *critical path* to be identified (ie the particular activities most critical to the overall time schedule). The network may be redrawn many times to show the effects that changes of resources and activities can have on the overall timing of the plan. It is also possible to discover, while the plan is being implemented, the effects on the overall plan of slippage or gains of time on individual activities.

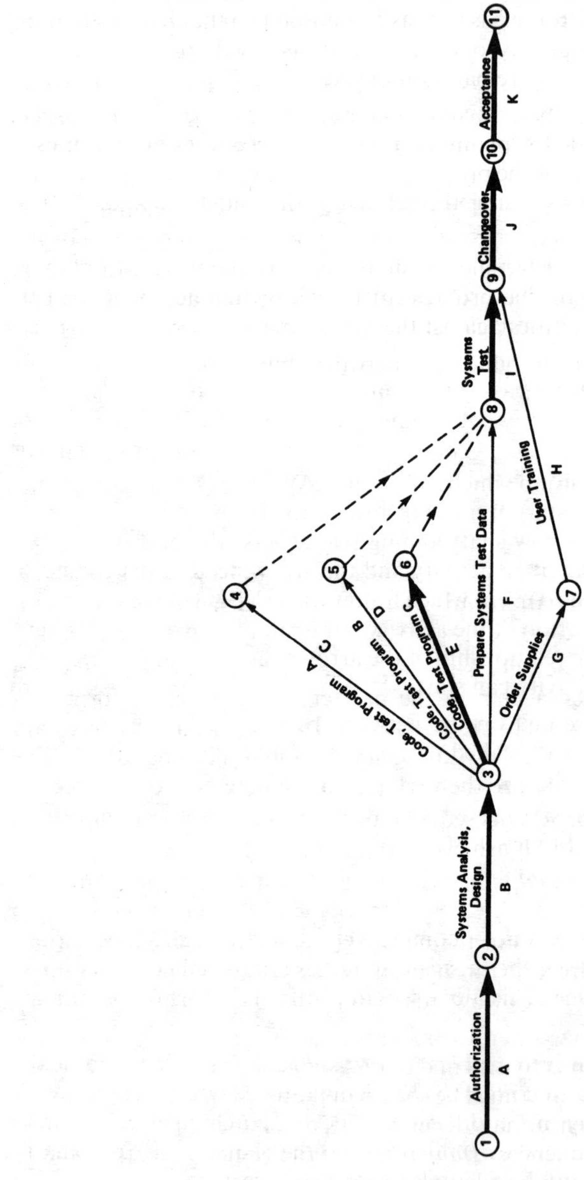

Figure 23.1 Network diagram

The limitation of networks is that the plan is not drawn to a time-sc
no simple inferences about timing can be made at a casual glance, as is t
case with bar charts. Progress is not easily recorded on the chart itself, nor is
the current position at any one time easily assessed from the network diagram, if any activities are part completed.

It follows that the best system is a mixture of both methods. The networks can be used: for planning, by showing the interrelationships of the activities analysed for the project; for scheduling, by assigning the times to each activity and assessing the critical path; and finally, for overall planning and control.

Bar charts are best used to show systems analysts the activities and target dates assigned to them, the progress of the individual activities, and the progress of all the activities against the overall plan.

It is not too difficult to build up a network, but there are certain terms which must be understood.

Activities and events

An *activity* is a task or job which takes time. An *event* is an instantaneous point representing the start or finish of an activity. Each activity in a project is represented by an arrow shaft joining two events; the butt of the shaft starts at the beginning of an activity and the arrow head at the end of the activity points at the next event which follows. The length of the arrow does not represent the duration of the activity but is simply a matter of drafting and convenience. The relationship of the arrows shows what must be done before a particular activity can start.

An event is represented on the network by a circle; the event at the beginning of the activity is called the *start* event and the one at the end the *finish* event.

It is usual to number events sequentially and to identify activities by the numbers of the start and finish events, eg 1–2, 2–3, 3–5 (fig. 23.2).

Dummy activities

Wherever possible, arrows in the network are drawn so that the flow of work is in the same direction throughout the network, either left-to-right or from the top of the sheet downwards.

Usually a project has several sequences of activities; there may be some relationships between activities in different sequences. These relationships are time-relationships and must be shown in the arrow network because an activity in one sequence may depend for its start on the completion of activities in other sequences. *Dummy* activities are needed to show these relationships; they neither utilise any resources nor consume time (except as below) and are drawn in the arrow networks as a broken line solely to

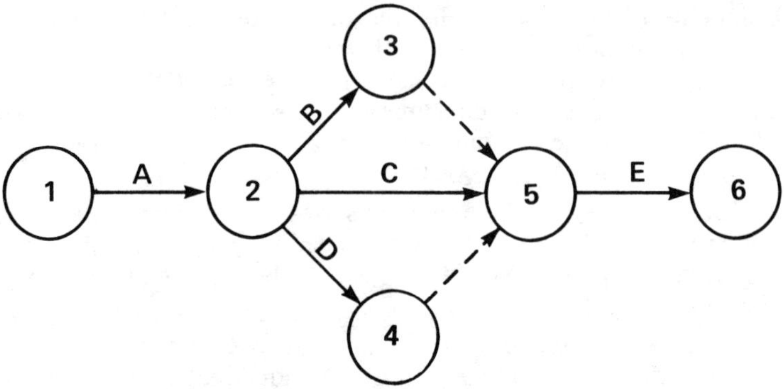

Figure 23.2 Network activities, events and dummy activities

identify the relationships. Each dummy must be directed to show which way the relationship connects the activities (fig. 23.2).

Ladders

A ladder describes a situation in which a number of activities overlap each other; in figure 23.3 the activities involved in data input are shown as

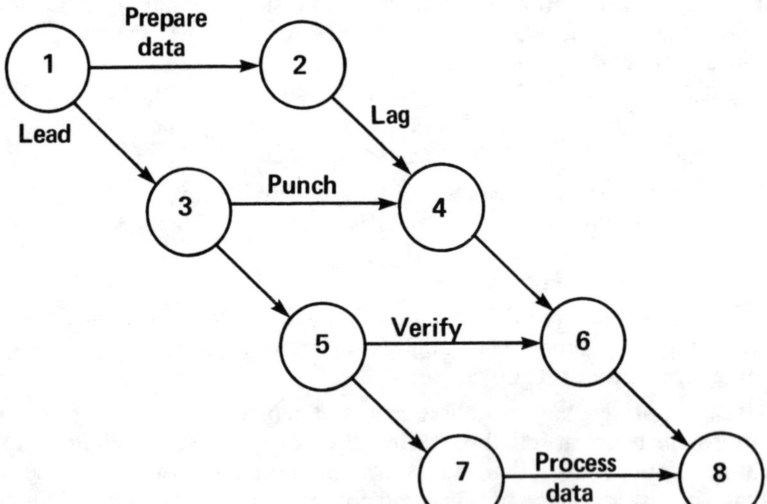

Figure 23.3 Network ladder

punching data, followed by verifying, followed by processing. Usually, however, verifying can start as soon as one batch of data has been punched, and processing as soon as a batch has been verified. To show this, arrows are used which indicate the lead between the start of one activity and the time to start the next one, and the lag between the finish of the first and the finish of the next activity. Lead time can be shown by using *real-time dummy* activities; these differ in that they require time, but like dummy activities do not use resources.

Network analysis

This is the process of giving times to each activity, finding the earliest start and latest finishing times of each event and then establishing the *critical path*. A critical path is the series of activities which determine the overall length of the project and which cannot be delayed without delaying the completion of the project (events 1–4–5–6 in fig. 23.4).

The first step is to enter the expected durations for each activity on the network. Then start at the earliest event and give it a notional time of zero to represent the beginning of the whole process. Working in the direction of the arrows, calculate the *earliest time* each event can happen by adding the time for the intervening activity to the earliest time for the previous event.

In the case where more than one activity leads into one event the earliest time for that event is the longest time taken by each of the activities. The total time of the network is the earliest time of the last event.

To calculate the *latest time* that each event can take place without delaying the time of completion of the last event, the procedure is reversed, working backwards from the last event and subtracting the time taken by the intervening activities. The earliest and latest time can thus be calculated cumulatively for each event and entered within each event symbol.

The difference between the earliest start and latest finish time can then also be recorded within the event symbol. Those events with zero difference are on the critical path (ie those events for which the timing is critical).

Floats

There are three useful values which can be derived from the earliest and latest times of events but which relate to the activities.

Total float is the amount of surplus time available on an activity before it affects the critical path. It is calculated as the latest time of the head event less the earliest time of the tail event, less the time necessary for the activity. The total float can be absorbed by a specific activity without affecting the critical path, but it can reduce the float of subsequent activities (for activity $E = 10 - 1 - 4 = 5$).

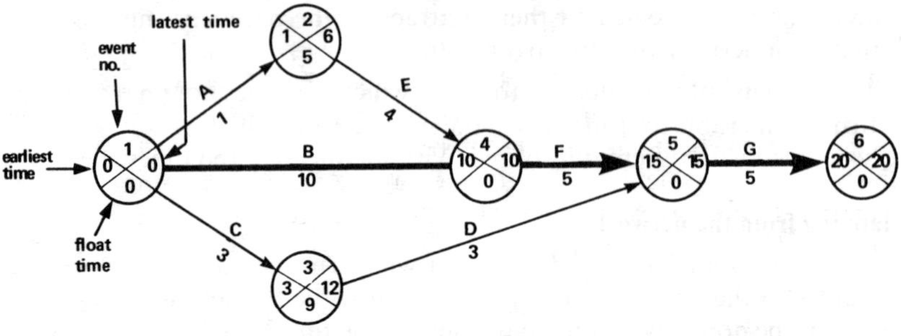

A	serve cereal	(1 min)
B	cook bacon and egg	(10 min)
C	boil water	(3 min)
D	make coffee	(3 min)
E	eat cereal	(4 min)
F	eat bacon	(5 min)
G	drink coffee	(5 min)

Activity	Event Nos	Earliest Start	Latest Finish	Float Time
A	1–2	1	6	5
B	1–4	10	10	0
C	1–3	3	12	9
D	3–5	6	15	9
E	2–4	5	10	5
F	4–5	15	15	0
G	5–6	20	20	0

Critical path is 1→4→5→6

Figure 23.4 Network analysis

Free float is the amount of free time which can be absorbed by an activity without reducing the float on subsequent activities. It is calculated as the difference between the latest finishing time of the head and tail events, less the necessary time for the activity (for activity E $= 10-6-4 = 0$).

Independent float is the amount of time by which an activity can expand without affecting any previous or subsequent activity. It is calculated by taking the latest starting time of the tail event from the earliest finishing time of the head event, and then subtracting the necessary time for the activity (for activity E = 10 − 6 − 4 = 0).

The main use of these floats is to assess which activities not on the critical path are vulnerable to stoppages in the schedule and which activities can be re-scheduled with different resources without affecting the timescale.

Planning from the network

After the network has been finished, it will form the basis for allocating resources to the activities. The plan must now be broken down to show what is to be done, by whom, and when, in the time-scale.

The problem arises of allocating resources to activities; if resources are limited, the duration of the project is likely to be extended. On the other hand, if time is short more resources may be required. If the use of resources is not planned with care, there will be overloads, with some activities delayed and others interrupted; or some resources will be idle for periods of time, with possible overloads later.

A bar chart and histogram, produced by a process of resource aggregation, provides an effective visual method of representation of resource requirements. Totals of resource requirements period by period show fluctuating demands which can be smoothed by rescheduling within the limitations of the plan. By adding the resources together it is possible to construct an aggregate profile of needs at the periodic times chosen. The float calculations can be used to show how the activities can be shifted in time within the limitations of the critical path to give a smoother allocation of resources.

Network packages

On a simple network, the arithmetic is not too laborious, but as the network increases in complexity, the calculations become more and more tedious and prone to error. When a network has more than a few hundred activities, it may pay to use a computer to do the calculations. Most computer manufacturers offer standard computer packages for this purpose. The best known of these are PERT (project evaluation and review technique) and CPM (critical path method).

When using a network package, the planner still has to identify the activities concerned in the project and to assign estimates of time to each activity. Where there is difficulty in estimating the times for activities a technique can be used whereby use is made of a weighted average of three estimates for each activity:

- optimistic time (A);
- most likely time (B);

– pessimistic time (C).

The formula used is the optimistic time multiplied by one, plus most likely time multiplied by four, plus pessimistic time multiplied by one, the total of which is divided by six $\left(\frac{A+4B+C}{6}\right)$

When networks are used for controlling projects as distinct from planning them, problems are encountered. It is not possible to report partial completion of an activity, posting only being possible on completion of an activity. Network packages also generate considerable amounts of information each time the network is updated. Added to this, many of the estimates of activity duration in a project network are likely to be amended a number of times during the life of a project. This means that the information generated by the package will not only be considerable but will keep changing; although this is merely a reflection of the current position of the network, it places a considerable burden on DP staff to absorb, interpret and act on the information generated. It is therefore essential that before using these packages the implications are fully understood and the necessary allowances and organisation changes made to cope with this situation.

Network – bar chart

This method sets out to combine the advantages of the box chart with those of the network by extending the bar chart to depict relationships between activities. Figure 23.6 gives a simple example of a network-bar chart which can be compared with the bar chart (fig. 23.5) and with the network (fig. 23.4).

For the network-bar chart, the lines representing each activity are drawn to scale in network sequence. At the end of each activity a short vertical time line is drawn to separate individual activities in the sequence. At any place where two or more activities must be completed before the next can be started, the vertical line is drawn at the end of that activity which is latent in time and is also extended to connect those activities which call short of this vertical line contain slack and this is shown by continuing the activity line in broken form. Thus in figure 23.5, activities A, E and B must be completed before F can be started and contain some float; and activities C, D and F must be completed before G can be started and contain some float.

Comparing this with the network diagram in figure 23.4 the network bar chart presents all the facts about the project but in an easier form than the network. Time values and float times are more easily seen, and the critical path is shown by those activities which form a continuous line through the chart.

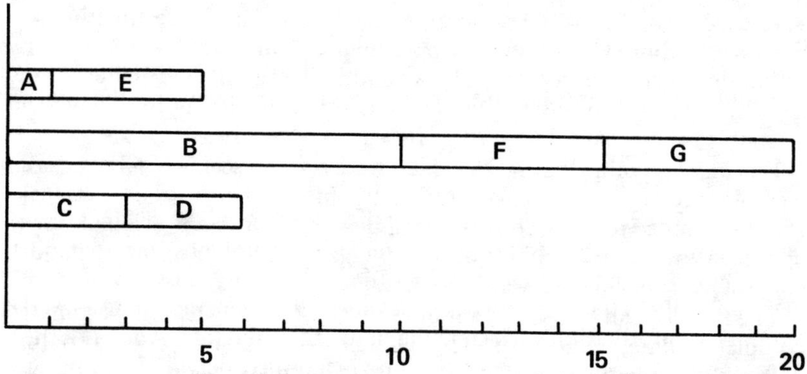

Figure 23.5 Bar chart

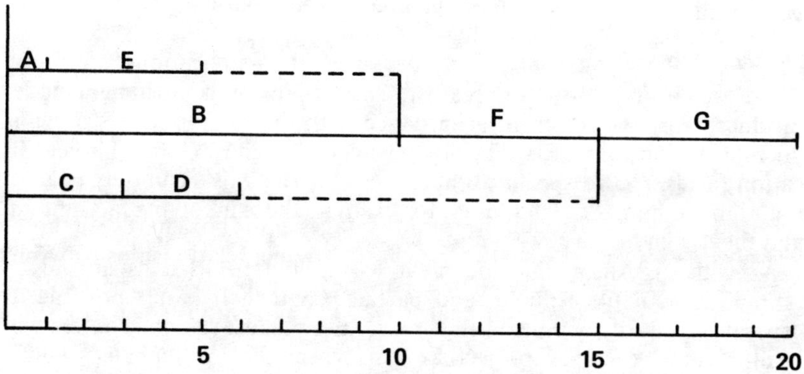

Figure 23.6 Network – bar chart

Checklists

Checklists can be used for several purposes and in a variety of ways.

Task lists

Task lists comprise a list of tasks associated with a given activity. The tasks are sub-divisions of an activity which are usually of small duration or at a low level. Standard task lists should be built up over the life of each project to produce as comprehensive a set of tasks as possible. When planning for a given activity in a particular project, the relevant tasks should be selected from the standard list.

Recorded performance should be used as an aid to estimating the resources required to perform a particular activity. This is not done by estimating each task in a formal way and building up an estimate for the activity as a whole, but by providing the estimator with more background to enable his estimates to be better informed.

It is quite possible that tasks, which were not foreseen when the activity was planned, will be revealed during the life of the project. Such tasks should be added to the list but physically separated from the original entries as a measure of the stability of a particular activity in terms of its task content over the life of several projects.

The task list can be used as a follow-up document by entering estimated completion dates against each task and checking (say, a week before that date) on the progress of that task. Provision should be made for insertion of the actual completion date if only as a means of ensuring that the task has been completed. Some estimate of partial completeness of the activity can be made by examining the task list to determine those tasks which have been completed, compared with those still to be done.

Document lists

These are another form of check list. They comprise the documents to be produced for an activity in accordance with the organisation's systems documentation standards. The documents include flow charts, file specification forms, record specification forms, etc. These lists can only be used for a limited number of activities, for example those involving fact-finding and the design of the new system.

As with task lists, the document list should be used as an aid to estimating activity resources and partial progress. It is not possible to generate standard document lists for activities because size and content will be different for each new project. For this reason the document lists will be added to during the life of the activity and estimates of partial completeness of the activity must be provisional.

It is not necessary to estimate times for the completion of individual documents, as it is unlikely that planning will be possible down to such a low level.

Responsibility lists

Responsibility lists are used for those parts of projects, especially if they are concerned with what is usually termed implementation, in which the responsibility for carrying out particular tasks is not always clear. For example, responsibility for ordering new equipment, training staff, preparing test data and transferring files, can rest with different departments in the organisation. The main purpose of a responsibility list is to assist the project leader by ensuring that responsibility for the completion of tasks has been allocated.

The list is usually prepared by the project leader and it consists of those tasks for which responsibility could lie outside the DP department. Those to be responsible for the tasks, and their departments, should be listed. The tasks which appear on the list will probably be the same for different projects within the same organisation; so, as with the task lists, a standard responsibility list can be developed within an organisation.

PROJECT CONTROL

The primary purpose of a control system is to provide the means whereby projects may be developed and completed within previously agreed times and budgeted resources. It must be possible to identify the status of development at any time in terms of resource usage against project progress; and to respond quickly to any deviations from planned performance. Its secondary purpose is to provide statistical information for overall data processing management, such as departmental costing, budgeting and performance statistics.

The control system must operate at levels and units compatible with the plans. It is recognised that planning, with some qualifications, represents ideal solutions which become progressively more realistic as development progresses. The control system must therefore be able to compare actual development with planned development and be capable of dealing with any deviations from those plans. Deviations must be expected in a development project because of the uncertainty inherent in forecasting the behaviour of future events. The extent of the variations will depend on the adequacy of the forecasting techniques used, the level of forecasting, and the extent to which the actual development conforms to the planning schedules.

A control system must therefore be capable of:

- measuring progress;
- reporting deviations;
- taking corrective action;
- reporting status and statistics.

Measuring Progress

Progress can be measured either on completion of activities or at time intervals, or both. Measurement on completion of an activity is associated with network methods of control, the degree of control exercised being a function of the size and duration of the activity specified. Measurement based on time intervals involves estimating partial completeness of an activity, since the end of an activity and the end of a time interval are unlikely to coincide.

The importance of time interval reporting, particularly from the senior management's point of view, is that it provides periodic re-evaluation of the project as a matter of routine. Without periodic reporting it is possible for senior management to know nothing of a project which has gone disastrously wrong until it is too late to take remedial action.

Recording progress

It is first necessary to collect information about current resource usage related to specific project activities. The most effective means of achieving this is by formally recording performance data on time sheets; verbal reports are not so effective, while they can assist with progress status identification they do not give an accurate indication of resource usage. The advantages of using time sheets as a basis for collecting project data are that:

- as formal reports they underline the importance of accurate performance reporting;
- they resolve a genuine difficulty in remembering time actually spent on the main project activities and other interrupting activities by facilitating daily entries for frequent interruptions or longer period entries for uninterrupted activities;
- they identify interruptions of main project work to management;
- they identify time spent on overhead activities;
- they form a basis for monitoring and controlling career development, staff appraisal, education and training;
- when used with a standard project and non-project coding structure, they can be used as input data to a manual or a computer-based control system.

The period covered by the time sheets and the unit of time recorded varies between installations according to the type of project and the conditions appertaining. However, weekly reports with a daily recording facility are commonly used.

While it is relatively easy to measure actual resource usage to date, it is more difficult to measure actual project development progress in such absolute terms. There are only two basic methods: ie by informal judgement or by monitoring actual development against predetermined planning stages.

The significance of breaking down projects to their component activities has been discussed for estimating and scheduling. The same reasoning applies to progress measurement:

- conformity to planned activities;
- ease of measuring clearly defined constituent parts;
- ease of measuring smaller components.

A coding structure can be devised which allows analysts to record their progress at hierarchical levels and their constituent activities: ie at project, phase, activity or milestone levels. The advantage of using hierarchical levels for progress measurement is that they are clearly identifiable on network diagrams and planning schedules, and allow progress to be identified against the background of the whole project.

Sometimes a status report is called for which involves the systems analyst making an estimate of the amount of progress in part-completed activities. The estimate is usually expressed as a percentage, based on the opinion of the systems analysts carrying out the activity. The task list discussed earlier can be also used to assist the systems analyst with estimating progress for part-completed activities. The activity is broken down into a number of small tasks; each task is allocated a percentage completion related to the whole activity which is allocated one hundred percent. The percentages of those tasks that have been completed when added up represent the percentage completion to date of the whole activity.

Deviations from the plan

Measurement may reveal that actual progress has fallen short of, or improved on, expected progress. The former result is more likely and is often referred to as slippage, but gains in time can also lead to inefficient use of resources. The term 'deviation' is used to cover both possible results. The effect of deviations on the remaining parts of the project will vary according to the extent and cause of the variances and their effect on the critical path.

The control system should calculate the effect of deviations in resource-time and elapsed time on the work remaining to be done. Some assumption must be made, for example, that the rate of progress to-date is likely to continue and may be projected over the remainder of the activity. This allows subsequent targets to be reviewed and decisions to be made on resource allocation, based upon performance to-date. It may be disputed that the rate of progress to-date is likely to continue, since there may have been special reasons which no longer apply; but there should be no dispute that a re-assessment is necessary.

When the project is reviewed, decisions will have to be made whether to reallocate resources or to reschedule activities. It is often tempting to assume, particularly in the early days of a project, that it will be possible to make up time later on; this hardly ever happens. It is more likely that escalation of costs will result in a watering down of the proposed system to keep within budgeted development costs, and potential benefits will be whittled away until the system no longer produces the proposed benefits.

Project reviews necessitate a ruthless attitude involving if necessary, stopping a project when the evaluation at review time indicates that the original justification does not now apply.

Performance statistics

Performance statistics collected by a control system can be of value in making estimates for subsequent projects or individual activities. In many computer installations performance data is notable for its absence, but even where it does exist it often proves of little value. This is largely because the control system for which the data is collected does not define the activities to be performed in a way which makes the data suitable for re-use. The use of performance data for resource and elapsed time estimating can only be of real value if definitions are consistent within an installation, and if the factors affecting the earlier estimates have been recorded.

It is important that computer installations continue to collect performance data and do not abandon the attempt because of the difficulties of defining the parameters.

The difficulty in providing a basis for the recording of performance data is that there must be a considerable degree of standardisation of these parameters before the data can be interpreted in a meaningful way to be used for future planning and estimating.

The activities against which performance is recorded must be defined and used consistently within an organisation. There are many factors which affect the performance of these activities; for example, taking systems analysts as a resource type, the extent of their technical and business experience will be reflected in their performance. These factors will vary with each project and between systems analysts. In addition, the nature of projects vary, one from another, according to their degree of complexity and innovatory content.

Tables can be devised which give numeric weightings to these factors; for example, technical and business experience, and the degree of innovation, difficulty or complexity contained in the activities. However, the calculation of these weights and their applicability to systems analysts must be influenced by subjective factors.

Finally it must be stressed that standard system documentation for procedures and data is essential for project planning and control.

SUMMARY

Project planning and control are essential in a system development situation to ensure good results. The high risk nature of computer projects demands that they are planned in the light of overall company strategy by a computer development steering committee. Planning and control requires the breaking down of a project into stages, each with a control point; the stages can be used for estimating timescales, allocating resources and measuring progress against the plan. Tools such as Gantt charts, Networks, Network Bar-charts, and Checklists can assist in the planning and control process. Control involves measurement of progress (usually via progress meetings and recording of work on timesheets), identification of deviations, taking corrective action and producing performance statistics. The careful planning and control of a computer project provides the right climate for success.

Appendix A
ESCOL Case Study

What follows is a reconstruction of a set of documents assembled during the investigation of an existing system. The case study first appeared in a slightly different form, in the NCC Mark 3 Basic Systems Course. The present version borrows some ideas developed subsequently by the Australian Department of Defence; these are gratefully acknowledged.

The documents are referenced and filed in accordance with the scheme published in the NCC Data Processing Documentation Standards. A key to the document reference is given (fig. 7.1 Vol. 1).

The case study allows the reader to try out skills taught in the body of the book. Suggested tasks are related to the chapters of the book, but course tutors can use the case study in any way which fits their course.

It should be noted that the documents do not provide a complete record of a system investigation, as that would be beyond the scope of this book. The reader will, therefore, have to make various assumptions where details are missing.

The computer configuration (reference PS/5.3/HARDWARE/1) is provided as an example. If the reader is familiar with another configuration appropriate to the problem situation, the known configuration may be substituted.

For various reasons no solutions are provided for the tasks. Firstly, there is no 'right' answer to any system design problem; any answer that meets the objectives and can be justified is acceptable. Secondly, the amount of documentation needed to provide a solution would be too large for this book. Thirdly, it is felt that the analysis and design tasks will best be carried out under supervision in a teaching situation where the tutor can provide an appropriate solution.

Title	System	Document	Name	Sheet
Background Information on the Educational Suppliers Co. Ltd. (ESCOL)	SOPS	1	General	1

The Company was first set up in 1946, by two ex-servicemen, John Curran and Cedric Watson, with the object of supplying educational materials in schools. Once established, they started producing various types of stationery for sale through retail outlets. This was followed by the production of boxed games, also for sale to retail shops.

In 1955, they bought their first retail store, selling stationery, children's games and toys and sports goods. The following year they set up a wholesaling operation for toys and sports equipment, and were able to use their existing connections with schools to sell sports equipment.

They now have six retail stores (in and around the city of Canbury), a printing works (also in the city), and a main warehouse just off the motorway, 30 km west of the city. All the stationery they sell, and the main contents of their boxed games, are produced at the printing works; the final assembly of the boxed games is done in a special department of the warehouse. All the sports goods are bought in either manufacturers or importers.

John Curran is now Managing Director and Cedric Watson is Sales Director.

The products are classified into:

- social stationery:
  in addition to writing pads and envelopes, this includes such things as wedding invitations and personal greetings cards;
- general stationery:
  this includes company letterheads, general office and typing paper and various single-part forms;
- educational:
  this includes exercise books, graph paper,

drawing paper and art materials. Most of these sales are made to local education authorities on a contract basis, but there is also a substantial volume of business with the retail trade and with educational wholesalers in other parts of the country;

- sports goods and games:
  the main market, consisting of the company's own stores and other retailers, is very seasonal, with 90 % of deliveries being made between October and December. The remainder of the market consists of schools and sporting bodies.

A small systems development team has just been set up, with the expectation that it would report to the Financial Director (Godfrey Shorthouse); but this was not acceptable to either the Sales Manager or the Production Manager (George Cox). To overcome the problem, a new post was created, that of Services Director, and the live-wire Personnel Manager Jerry Wickens, promoted to this. His deputy, Marian Ryan, succeeded him as Personnel Manager. For the present, the systems team reports direct to Mr. Wickens.

Title	System	Document	Name	Sheet
Organisation Chart	SOPS	1	ORG	1

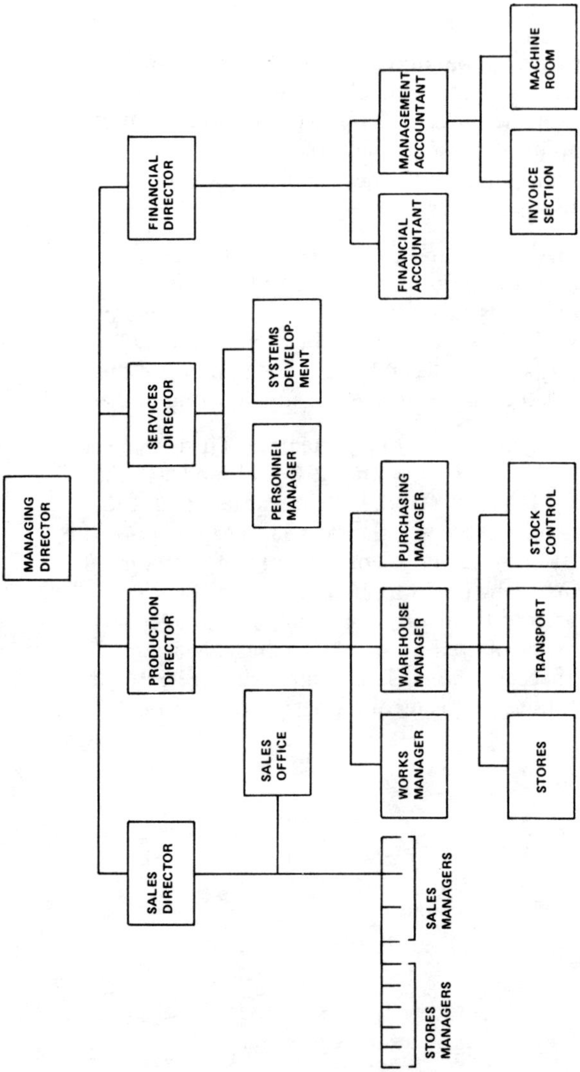

Title	System	Document	Name	Sheet
Terms of Reference	SOPS	1	TOR	1

The company requires a suitable system from receipt of order through invoicing, stock control and sales ledger. Considerable stress has been placed on the need to improve the flow of information to management; product categories, profitability reporting and debt collection monitoring are the major deficiency areas in the existing system. The stock ledger system is often out-of-date, and it is very difficult to extract information from it.

The system must be capable of producing the following outputs:

- invoice set;
- statement;
- stock list;
- stock valuation;
- price list;
- price changes;
- analysis of fast/slow-moving items;
- stock action report (items breaking maximum, minimum or re-order levels);
- sales analysis (by salesman, area, department or outlet);
- credit list (accounts exceeding credit limit);
- age of debt analysis;
- inactive customers;
- labels for direct mail.

Educational product group will not be included in the system initially, but provision should be made to include this once the system is running successfully with the other product groups.

The system should allow any order for which all items are in stock to be despatched the day after receipt of order.

Title	System	Document	Name	Sheet
Invoice Section	SOPS	2.1	INVSECT	1

Participants	Date
M.O.R. Cash (Supervisor) S. Williams (Systems)	5.2.78
	Location Invoice Office
Objective Agenda To establish invoicing procedure	Duration 50 mins.

Results:	Cross-reference

Orders are received from the machine room and a four-part invoice set is typed as follows:

Top copy (white) is the customer's invoice copy but at this stage is retained with copy two until the goods are despatched. 4.1/INV

Copy two (blue) is the accounts copy and remains with the top copy until the goods are despatched. It is then used as source medium for posting to the sales ledger and is eventually filed in invoice section.

Copy three (yellow) is the warehouse copy and is their authority to make up the order and despatch to the customer. The pricing details do not appear. This copy is returned to invoice section when the goods are despatched.

Copy four (green) is the delivery note. It carries similar details to copy three and is enclosed with the goods when they are despatched to the customer.

On receipt of the warehouse copy, the top copy is forwarded to the customer and the accounts copy is batched before being passed to the machine room with the pre-list control total for updating the sales ledger. The same procedure is followed with the warehouse copy which is used for updating the stock ledger.

STAFF

Supervisor and three typists/filing clerks.

Title	System	Document	Name	Sheet
Machine Room	SOPS	2.1	MACHROOM	1/2

Participants	Date
R. Ferguson (Machine Room Supervisor) S. Williams (Systems)	3.2.78
	Location Machine Room
Objective/Agenda To ascertain machine room procedures and problems	Duration 35 mins.

Results: | Cross-reference

STAFF

Supervisor and three operators.

EQUIPMENT

Three accounting machines (13, 10 and 4 registers) and two add-listers.

ORDER EXTENSION

Orders are received in batches from the Sales Office for pricing, extension and calculation of discounts and sales tax where applicable. No sales tax is payable on orders from schools or education authorities where the order is certified by the originator 'For Educational Purposes'. Orders are then passed to the Invoice Section for preparation of invoices.

SALES LEDGER

There are 20,000 customer accounts (of which, it is estimated, only 25% are active) divided into geographical areas.

Posting of invoice details is from the blue copy of the invoice and the statement and ledger card are updated simultaneously. The posting documents are batched and pre-listed, and at the end of each batch the control total is checked with the machine total on the posting summary. Normal control accounts are maintained. Cash and returns are posted in a similar manner. At the end of the month statements are extracted from the trays and forwarded to customers.

Cross-reference: 4.1 SALEDGR; 4.1 STATMT; 4.1 SALEDGRPS; 4.1 SALEDGRCA

Title	System	Document	Name	Sheet
Machine Room (cont.)	SOPS	2.1	MACHROOM	2/2

Participants	Date
	Location
Objective/Agenda	
	Duration

Results: | Cross-reference

STOCK LEDGER

There are 5,500 items divided as follows:

- social stationery: 600 (3 product categories);
- general stationery: 2,300 (6 product categories):
- educational: 1,800 (4 product categories);
- sports goods and games: 800 (12 product categories).

Issues are posted from the warehouse copy (yellow) of the invoice. The posting documents are batched and pre-listed before posting to the ledger, and at the end of each batch the control total is checked with the machine total on the stock ledger summary. For each movement a line appears on the Stock Control Report and from this is produced the Stock Report Action Slip calling for re-order when the free stock falls below re-order level or urge (ie, chase up orders) when actual stock falls below re-order level. Purchase orders, appropriations and returns are treated in a similar way.

Cross-reference: 4.1 STOCLED; 4.1 STOCLEDSUM; 4.1 STOCREPT; 4.1 STOCREPTAS

Goods inwards (including receipts and returns) are posted before goods out. The number of goods inwards notes per day would average 100.

Title	System	Document	Name	Sheet
Order Entry System	SOPS	2.1	ORDENT	1

Participants L. Smith (Sales Office Manager) S. Williams (Systems)	Date 3.2.78
	Location Sales Office
Objective Agenda To ascertain function of Sales Office and any problems	Duration 45 mins.

Results: | Cross-reference

Orders average 200 per day, 6 items per order. In pre-season rush, average 300 per day, 10 items per order. About half received from rep or store on standard order form, others direct from customer by post or 'phone to be transcribed to standard form by sales clerks.

All orders vetted for correctness of item description and complete delivery instructions. Credit checked against 'black-list' provided weekly by Accounts. Queries referred to Sales Manager.

Orders then passed to Stock Control to check on availability. Then returned to Sales Office and order acknowledgement sent with copy to salesman.

Orders then passed to Machine Room for pricing, extension and calculation of discount and sales tax.

PROBLEMS

With telephoned orders, customers usually want to know availability. Then necessary to 'phone Stock Control either with customer hanging on or ringing customer back later. This holds up normal work of order-vetting. During December 80% of orders received by telephone.

In a normal week 60% of orders received on Monday and Tuesday; overtime normally worked on these two days.

Sales Managers require sales report for previous week by Tuesday mid-day (compiled from the orders).

STAFF

Manager, six sales clerks, two typists.

Title	System	Document	Name	Sheet
Sales Management	SOPS	2.1	SALES DIR	1

Participants	Date
C. Watson (Sales Director) S. Williams (Systems)	7.2.78
	Location CW Office
Objective/Agenda To establish information required by Sales Director and any problems	Duration 25 mins.

Results: | Cross-reference

Responsible for overall marketing policy. Has four sales managers for Products Division and six store managers.

REPORTS RECEIVED

Monthly sales – by category
– by outlet
– by representative

(each sales manager receives his appropriate part of these reports)

Weekly – total value of wholesale orders
– sales of each retail store

PROBLEMS

Sales of general stationery are dropping. Believed to be due partly to competition from local small printers and partly from larger printers, set up to produce multi-part and continuous stationery. Considering reducing the range of this group and merging remainder with social and educational.

Monthly sales reports are always late – sometimes by as much as two weeks.

Title	System	Document	Name	Sheet
Warehouse	SOPS	2.1	WHSE	1

Participants	Date
H. Black (Warehouse Manager) S. Williams (Systems)	4.2.78
	Location Warehouse
Objective/Agenda To discover any problems in existing system	Duration 1 hour

Results:	Cross-reference
Many orders have to be returned to Sales Office because item not in stock. (Estimated 5%) Shortage of warehouse space as stocks of toys and boxed games build up from May to September. Normal duties of warehousemen frequently interrupted by calls from Sales Office to check whether an item is physically in stock.	

Title	System	Document	Name	Sheet
Proposed Configuration	SOPS	5.3	HARDWARE	1

1 central processor
1 operator's console
2 disk drives
2 terminals with CRT display (VDU)
2 terminals with CRT display and low speed character printers
1 line printer

Central processor

Has a memory of 48K words, each word being 16 bits in length and hence capable of storing two 8-bit characters. Memory cycle time is 1 microsecond.

Operator's console

Consists of a standard keyboard and low-speed character printer (10 characters per second) and is used mainly for communication between operator and operating system.

Disk drives

Each has a single spindle with moving head and removable 10-surface disk pack (data capacity of each 20 million characters). Access time 80 milliseconds.

Terminals

All four terminals are connected to CPU via a multiplexor. Terminals may be interfaced with standard PO lines. Two of the four have also a keyboard, with low-speed character printer (width 70 characters).

Line printer

Can accept a variety of pre-printed stationery. Printing speed is 1,200 lines/minute and maximum width is 132 characters.

Title	System	Document	Name	Sheet
Invoice	SOPS	4.1	INV	1

INVOICE

THE EDUCATIONAL SUPPLIES COMPANY LTD.
121 HIGH STREET
CANBURY

Tel: 705335 Telegrams: TESL

Invoice Address .... Delivery Address ....

Your Order No.	Date	Our Order No.	Delivery per

Part No.	Description	Qty. Ord.	Qty. Del.	To Follow	Price/ Unit	Value

Total Value	
− Discount	%
+ Sales Tax	%
Amount Due	

Note: Top copy (customer's) and copy two (accounts) are identical as above but different colours; copy three (warehouse) and copy four (delivery note) are identical (but different colours) and do not include price, discount price, sales tax, or value.

Title	System	Document	Name	Sheet
Sales Ledger	SOPS	4.1	SALEDGR	1

SALES LEDGER

SHEET No. : 13

Name: CUNNINGHAM & CO. LTD.

Address: WEYMOUTH ROAD
CANNINGTON

Date	Details		Debit	Credit	Balance	Verification
4 JAN	GDS	40.667	54.77		54.77	109.54
6 JAN	GDS	40.714	102.68		157.45	314.90
6 JAN	GDS	40.787	34.21		191.66	383.32
15 JAN	GDS	40.854	4.50		196.16	392.32
18 JAN	CSH	1.022		153.52		
18 JAN	DIS	1.022		3.92	38.72	77.44
25 JAN	GDS	40.965	50.76		89.48	178.96
1 FEB	GDS	41.214	24.63		114.11	228.22
2 FEB	GDS	41.296	10.14			
2 FEB	GDS	41.297	8.55		132.80	285.60
5 FEB	RET	425		4.64	128.16	256.32
16 FEB	GDS	41.737	100.40		228.56	457.12
24 FEB	CSH	2.150		37.87		
24 FEB	DIS	2.150		0.84	189.85	379.70
24 FEB	GDS	42.120	135.00		324.85	649.70

Title	System	Document	Name	Sheet
Sales Ledger Control Account	SOPS	4.1	SALEDGRCA	1

SALES LEDGER

SHEET No. : 47

Name: CONTROL ACCOUNT

Address: A TO C

Date	Details	Debit	Credit	Balance	Verification
16 FEB	B/FWD			3.517.52	7.035.04
16 FEB	GDS	496.61		4.014.13	8.028.26
16 FEB	CSH		241.38		
16 FEB	DIS		8.79	3.763.96	7.527.92
17 FEB	GDS	281.75		4.045.71	8.091.42
17 FEB	CSH		301.83		
17 FEB	DIS		14.16	3.729.72	7.459.44
18 FEB	GDS	591.16		4.320.88	8.641.76
18 FEB	RET		12.33	4.308.55	8.617.10
18 FEB	CSH		205.15		
18 FEB	DIS		10.20	4.093.20	8.186.40
19 FEB	GDS	376.24		4.460.44	8.938.88
19 FEB	CSH		131.44		
19 FEB	DIS		8.18	4.329.62	8.659.64
22 FEB	GDS	280.81		4.610.63	9.221.26
22 FEB	CSH		109.30		
22 FEB	DIS		9.42	4.491.91	8.983.82
23 FEB	GDS	482.96		4.974.87	9.949.74
23 FEB	RET		15.10	4.959.77	9.919.54
23 FEB	CSH		180.17		
23 FEB	DIS		12.17	4.767.43	9.534.85
24 FEB	GDS	368.77		5.135.20	10.272.40

Title	System	Document	Name	Sheet
Sales Ledger Posting Summary	SOPS	4.1	SALEDGRPS	1

POSTING SUMMARY

SALES LEDGER
Section A TO C

SHEET No. : 178

O/d Balance Proof	Date	Details	Debit	Credit	Balance	Verification
			0.00		0.00	0.00
0.00	24 FEB	GDS 42.084	6.53		48.21	98.42
0.00	24 FEB	GDS 42.079	8.77		14.54	29.08
0.00	24 FEB	GDS 42.076	26.54		46.74	93.48
0.00	24 FEB	GDS 42.075	5.62		11.67	23.74
0.00	24 FEB	GDS 42.080	3.68		6.94 cr	13.69 cr
0.00	24 FEB	GDS 42.088	21.12		56.95	113.90
0.00	24 FEB	GDS 42.082	2.64		22.90	45.80
0.00	24 FEB	GDS 42.121	48.32		90.53	181.06
0 00	24 FEB	GDS 42.136	50.00			
	24 FEB	GDS 42.137	60.55		211.17	422.34
0.00	24 FEB	GDS 42.120	135.00		324.85	649.70
0.00	24 FEB	GDS	368.77		5.136.20	10.272.40

Title	System	Document	Name	Sheet
Statement	SOPS	4.1	STATMT	1

Please return this tear-off portion with your remittance

CUNNINGHAM & CO. LTD.
WEYMOUTH ROAD
CANNINGTON

Balance per Statement £

Adjustments

Amount of remittance

STATEMENT The Educational Supplies Company Ltd.,
121 High Street,
Canbury

Date	Details		Debit	Credit	Balance
				Balance from last statement	89.48
1 FEB	GDS	41.214	24.63		114.11
2 FEB	GDS	41.296	10.14		
2 FEB	GDS	41.297	8.55		132.80
5 FEB	RET	425		4.64	128.18
16 FEB	GDS	41.737	100.40		228.56
24 FEB	CSH	2.150		37.87	
24 FEB	DIS	2.150		0.84	
24 FEB	GDS	42.120	135.00		324.85
			368.77		
					The last amount in this column is now due

PLEASE NOTE: Cheques should be made payable to The Educational Supplies Company Ltd.

Title	System	Document	Name	Sheet
Stock Ledger Card	SOPS	4.1	STOCLED	1

STOCK RECORD

Account No. 1226 Sheet No. : 10

Part No. 1226

Description: WEDDING INVITATIONS
TYPE 6

Drawing No:

Unit of quantities
GROSS

Re-order 400
Max. 2000
Min. 200

Proof	Date	Details	Orders	Receipts	Appropriations	Issues	Balances				Usage to date	Check Total
							Orders	Stock	App.	Free		
0	9 MAR	B/FWD					125	750	140	735	789	3.470
	19 MAR	1.586			150							
	19 MAR	1.595			200		125	750	490	385	789	3.820
0	3 APR	7.846	550				675	750	490	935	789	4.370
0	13 APR	202				52	675	698	439	935	841	4.318
0	21 APR	354				56	675	642	382	935	897	4.262
0	2 MAY	1.123		125			550	767	382	935	897	4.262
0	6 MAY	600				73	550	694	309	935	970	4.189
0	11 MAY	845				79	550	615	230	935	1.049	4.110

Title	System	Document	Name	Sheet
Stock Ledger Summary	SOPS	4.1	STOCLEDSUM	1

STOCK SUMMARY

Sheet No. : 152

Proof	Date	Details	Orders	Receipts	Appropriations	Issues
0	11 MAY	723				18
0	11 MAY	1.011				17
0	11 MAY	387				17
0	11 MAY	104				27
0	11 MAY	845				79
						158

Title	System	Document	Name	Sheet
Stock Control Report	SOPS	4.1	STOCREPT	1

STOCK CONTROL REPORT

Sheet No. : 152

Proof	Date	Details	Orders	Receipts	Appro-priations	Issues	Balances				Usage to Date	Check Total	Re-order Level	Account No.	Signal	
							Orders	Stock	App.	Free					Re-order	Urge
0	11 MAY	723				18		500	120	380	152	2.366	400	1.214	20	
0	11 MAY	1.011				17	1.000	468	868	600	254	4.313	500	1.223		32
0	11 MAY	387				17	200	835	542	493	105	3.281	350	1.249		
0	11 MAY	104				27	400	293	415	278	680	3.343	300	1.255	22	7
0	11 MAY	845				79	550	615	230	935	1.049	4.110	400	1.266		
						158										

Title	System	Document	Name	Sheet
Action Slip	SOPS	4.1	STOCREDTAS	1

ACTION SLIP

Serial No. : 1001

Date	Account No.	Recorder Level	Recorder Signal	Urge Delivery
19 MAR	1.266	400	15	

TASKS

The tasks are presented in system development sequence rather than chapter sequence but each set of tasks is related to particular chapters.

Chapter 1

1. Identify the subsystems and interfaces of the ESCOL organisation as a whole.
2. Study the details of the sales order processing system and draw a diagram showing its subsystems and their interfaces.
3. Suggest some negative feedback mechanisms which will apply in the sales order processing system.
4. Comment on the organisation structure of ESCOL and any organisational problems which emerge from the details given.

Chapter 4

5. Select one of the approaches to system development which you think is appropriate to the terms of reference and explain why you have chosen it.
6. Formulate a plan for the further development of the ESCOL system which takes into account user involvement.
7. Analyse the benefits that are likely to accrue to ESCOL management from the proposed order processing system.

Chapter 5

8. Identify any extra information which you feel you need in order to design a system for ESCOL, and draw attention to any problems of information flow in the existing system.

Chapter 6

9. Compile a set of questions that you would wish to pursue in interviews with:
   - Financial Director;
   - Stock Controller;
   - Personnel Manager.
10. Describe what sort of things you would be looking for if you spent some time in the machine room to observe procedures.

Chapter 7

11. Produce a clerical procedure flowchart for the sales order processing procedure of ESCOL (drawn by department).

12. Draw a decision-table for the production of the stock report action slip (SOPS/4.1/STOCREPTAS).
13. Construct a document/item grid for all the sample documents (SOPS/4.1/—).
14. Complete a clerical document specification for the invoice (SOPS/4.1/INV).

Chapter 8

15. Define the objectives of the ESCOL sales order processing system.
16. Construct a decision/information grid-chart for the system.
17. Using 16, construct a recipient/information grid-chart, indicating by a code the frequency of receipt required.
18. Using 17, list the reports required for the system with their frequency.
19. Using 18, construct an input/output grid-chart showing all the data items required for the various output reports.

Chapter 9

20. Using 19, break the input data items down into items to be input each time, items to be calculated, and items to be stored.
21. Using the list of stored data items, carry out a TNF analysis.
22. Produce a logical data structure diagram for the records which have been normalised.
23. Identify the logical processes required to produce the outputs from the inputs and stored data, and indicate the nature of the processing.
24. Produce a system outline for the system and agree it.

Chapters 10, 12, 13, 14, 16

25. Using the data structure diagram, combine the logical records into physical records appropriate to the processing requirements, and specify these records.
26. Combine the physical records into files, and specify the files.
27. Select appropriate file organisation and access methods.
28. Identify the output media required for each output and produce an appropriate layout (for printer or VDU) except for the invoice and statement.
29. Design a dialogue (VDU) for entry of each item of input data for the system and specify the dialogue on a display layout.

30. Produce a system flowchart for the overall system.
31. Produce a computer min-chart for the computer procedures.
32. Produce an interactive system flowchart for the interactive procedures.
33. Produce an outline computer procedure flowchart for the non-interactive procedures.
34. Determine the security requirements for the system.
35. Produce detailed validation requirements for the validation procedures.
36. Identify the file reconstruction procedures, including back-up.
37. Design the controls for the system, including file controls.
38. Suggest methods of controlling access to the files.
39. Iterate 25 through 38 until an acceptable computer subsystem design is achieved.

Chapter 15

40. Design forms for capturing the sales order data, the cash received data, the goods received/returned data, and physical stock count data.
41. Design output computer documents on printer layouts for the invoice set and the statement, and complete appropriate computer document specifications.

Chapter 17

42. Design a product code and a customer code to include check digit facilities.
43. Review all codes in the system.

Chapter 18

44. Define and design the user procedures for sales order processing using the computer.
45. Produce a clerical procedure flowchart for all activities.
46. Recommend staffing levels and an organisation structure for the new system.

Chapter 22

47. Review the design carried out in tasks 25 to 46 and iterate until an acceptable, consistent system specification is available.
48. Produce a user manual for the system.

Chapter 19

49. Produce an outline of a training course for:
    - line managers;
    - clerical staff;

    in all departments affected by the new system.
50. Define test data to test the system.

Chapter 20

51. Define procedures for converting the sales ledger and stock ledger to computer media.
52. Choose a method of changeover and explain why it has been chosen.

Chapter 21

53. Produce a report on the quality controls of the system answering the appropriate questions in the checklist in this chapter.

Chapter 3

54. Produce a report to management explaining the new system in simple terms and describing the benefits that will accrue from adopting the system.
55. Produce an outline of a presentation of the system to the board of directors, including examples of the usual aids which you would use.

Appendix B
CODASYL Recommendations

INTRODUCTION

CODASYL is an international voluntary organisation, predominantly of users of computer systems, and represents a broad spectrum of interest in computing practice. Under its executive committee, there are a number of other committees including:

- Programming Languages Committee;
- Data Description Language Committee;
- Systems Committee.

To each of these committees report a number of Task Groups, the most significant being the Data Base Task Group (DBTG) which reports to the Programming Languages Committee.

In a number of detailed reports, the CODASYL DBTG has proposed standards covering the methods used to describe data stored in an on-line database, and the methods used to access and process data in an on-line database. These are two of a number of interfaces between a DBMS, its users and implementors. Figure B1 shows the overall relationships between components of database systems as recommended by CODASYL DBTG. The major components are:

- Schema Data Definition Language (DDL);
- Subschema DDL;
- Data Manipulation Language (DML);
- Device Media Control Language (DMCL);
- Data Base Management System (DBMS).

DATA DEFINITION LANGUAGE (DDL)

Both the Schema and the Subschema DDL is used to describe data structures in a database. The Schema DDL is used to describe the overall structure, and this description is maintained in a direct-access SCHEMA file which is used as a reference by other system components. The Subschema DDL is similar to the Data Division of a COBOL program in that the data for a particular program is described. The Subschema DDL allows the user to derive logical data requirements as a subset of the overall schema. The exact specification of DDL is the current work of the Data Definition Language Committee of CODASYL.

DATA MANIPULATION LANGUAGE (DML)

This language is defined by CODASYL DBTG as an extension to COBOL, and similar extensions to other established high-level programming languages are possible. DML is the language with which application programs communicate with the Data Base Management System. It is a 'host' language, in that the DML verbs used to access and control data in the database are freely mixed with conventional high-level language statements when the program is written.

DEVICE MEDIA CONTROL LANGUAGE (DMCL)

This is the language which specifies the allocation and mapping of data to secondary storage devices. Each type of hardware may have a separate set of DMCL, and research into these areas is being conducted by the 'Storage Structure Description Language Task Group' of the Systems Committee of CODASYL.

DATA BASE MANAGEMENT SYSTEM (DBMS)

The software routines which control the loading and running of a database system are sometimes called Data Manipulation Routines. The Data Base Management System consists of these as well as the other software components of the Data Base System which will go together to become the control software 'package'.

The systems analyst designs the database and the schema using the DMCL to implement the database and the DDL to implement the schema.

Subsequently, application systems are designed in the normal way except that a subschema must also be designed with each particular application.

Then, with the aid of DML and DDL preprocessors and a COBOL compiler, the source language instructions are compiled into a complete user application object program.

When the application is being run and a data request appears in the program, it is passed to the DBMS software resident in main storage; this refers to the schema and transfers the data requested from the database to the work area in main storage. The database can also be updated with data in the work area by the user application program via DBMS and the schema.

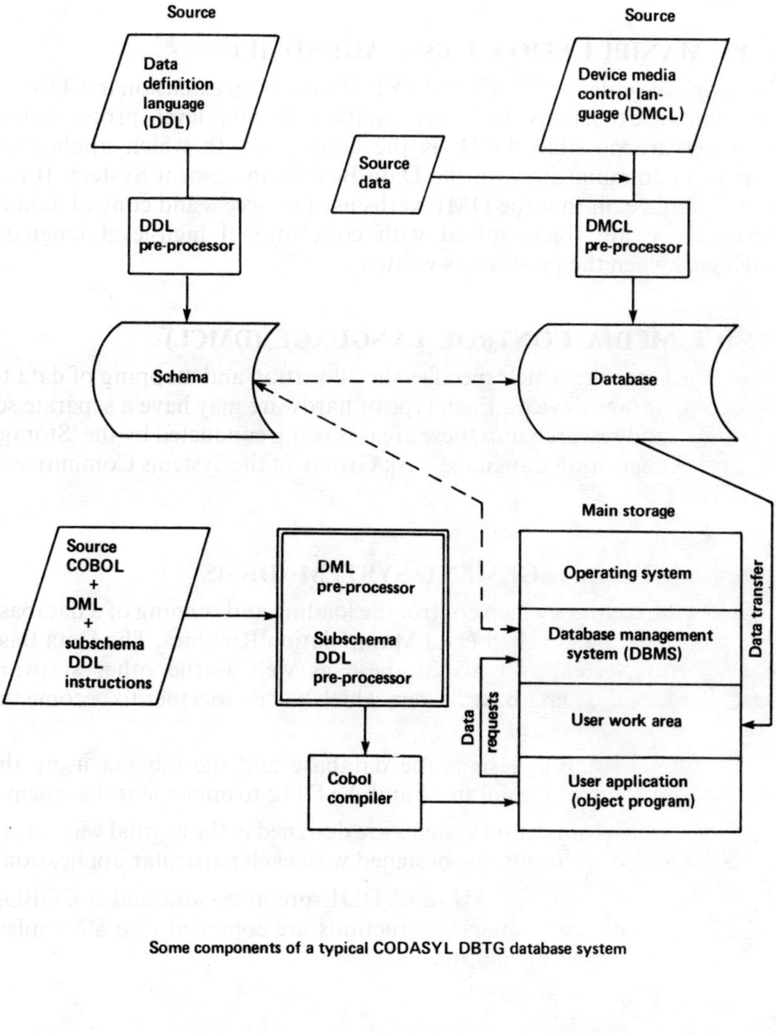

Some components of a typical CODASYL DBTG database system

Appendix C
Further Reading

The list below provides a guide for further reading on topics covered by this book.

Benwell N J (ed), *Data Preparation Techniques*, Advance Publications, 1976.

Brandon D H, Palley A D and O'Reilly A M, *Data Processing Management*, MacMillan, 1975

Chadwick B, Farr M A L and Wong K K, *Security for Computer Systems*, NCC Publications, 1972

Churchman C W, *The Systems Approach*, Delacorte Press, 1968

Civil Service Department, *Design of Forms in Government Departments*, HMSO, 1972

Collin W G, *Introducing Computer Programming*, NCC Publications, 1974

Couger J D and Knapp R W (ed), *Systems Analysis Techniques*, Wiley, 1974

Data Processing Documentation Standards, NCC Publications, 1977

Date C J, *An Introduction to Database Systems*, Addison-Wesley, 1976

Davis G B, *Management Information Systems*, McGraw-Hill, 1974

Gibbons T K, *Integrity and Recovery in Computer Systems*, NCC Publications, 1976

Gildersleeve J R, *Organising and Documenting Data Processing Information*, Hayden/NCC, 1977

Grindley K, *Systematics*, McGraw-Hill, 1975

Hartman W, Matthes H and Proeme A, *Information Systems Handbook*, Kluwer-Harrap, 1972

Lefkowitz D, *Data Management for On-line Systems*, NCC Publications, 1974

London K R, *Techniques for Direct Access*, Auerbach, 1973

London K R, *The People Side of Systems*, McGraw-Hill, 1976

Lowe C W, *Critical Path Analysis by Bar Chart*, Business Books, 1969

Martin J, *Computer Data-base Organisation*, Prentice-Hall, 1975
Martin J, *Design of Man-computer Dialogues*, Prentice-Hall, 1973
Milward G E (ed), *Organisation and Methods*, MacMillan, 1967
Moser C A and Kalton G, *Survey Methods in Social Investigation*, Heinemann, 1971
Mumford E, *Job Satisfaction – a Study of Computer Specialists*, Longman, 1972
Mumford E, *Systems Design for People* – Book 3 of *Economic Evaluation of Computer-based Systems*, NCC Publications, 1971
Mumford E and Sackman H (ed), *Human Choice and Computers*, North-Holland, 1975
Sanders D H, *Computers in Business: an Introduction*, McGraw-Hill, 1975
Sharratt J R, *Data Control Guidelines*, NCC Publications, 1974
Stamper R K, *Information in Business and Administrative Systems*, Batsford, 1973
Tucker R I, *Management Information and Control Systems*, Wiley-Interscience, 1976
Ward T B, *Computer Organisation, Personnel and Control*, Longman, 1973
Waters S J, *Introduction to Computer Systems Design*, NCC Publications, 1974
Woodgate H S, *Planning by Network*, Business Books, 1977
Wright G G and Evans D, *Commercial Computer Programming*, NCC Publications, 1975
Yourdon E, *Design of On-line Computer Systems*, Prentice-Hall, 1972
Yourdon E, *Techniques of Program Structure and Design*, Prentice-Hall, 1975

Appendix D
Glossary

GLOSSARY OF TERMS WHICH ARE NOT DEFINED IN THE TEXT

Access (v)

To access a data storage location or data: to carry out the actions necessary to be in a position to read or write the data from or to its storage location. *Direct Access:* the process of accessing a data item or record directly without the need to search through or read other data items or records. To go directly to the data item's or record's location. *Random Access:* synonym for direct access (USA).

Access Time

The time that elapses between the moment the command to access a location or area is given, and the moment when the transfer of data to or from that area can commence.

Activity

1. An operation comprising the time and resources needed to complete part of a project network, ie the passage of time and employment of resources that are necessary to progress from one event to the next.
2. A measure of the proportion of the records in a file that are processed during one run (ie file activity).

Address

An identification for a register, location in storage, or other data source or destination; the identification may be a name, label or number. An *Absolute Address* is the actual numeric identity in machine language of a fixed position in main store; distinct from *Relative Address*, which is the position of data in relation to some datum point, such as the first instruction in a program segment.

Address Generation

A method of obtaining the address of a piece of data by generating the address from known components of the data, usually involving a mathematical process.

Algorithm

A prescribed set of well-defined rules or processes for the solution of a problem in a series of steps.

Application Package

Generalised programs written for a major application area (eg payroll or vehicle scheduling) in such a way that they can be used by many organisations.

Area, Input or Output

A block of main storage reserved for data being transferred from and to the input and output units.

Assembler

A program which translates a source program written in symbolic language into machine or object language.

Audit Trail

The path of events from output back to original source data which auditors need to be able to trace.

AUTOSATE

*A*utomated data *S*ystems *A*nalysis *T*echnique – designed in the 1960s as a tool for documenting information systems.

Back-up

A compatible reserve, in the form of hardware, program, or file/record duplication, for use in case of breakdown.

Backing Store

A large capacity data store supplementing the main store and accessible by CPU command.

Badge Reader

A device for collecting data recorded as holes in pre-punched cards or on plastic badges.

Bar Code

Characters recorded in a coded bar shape.

Batch (v)

To group input documents or records so that they are dealt with together.

Batch Processing

An approach to data processing where similar input items are grouped for processing during the same machine run. Contrast with *real time processing* where data items are processed as they arise.

Batch Total

A total of some common component of a batch of data so as to enable a control to be maintained over the validity of the data. An example of this is the totalling of the cash value of daily receipts records so that at any time during the processing the cash total of the batch can be taken and related back to the original to ensure that no distortion of the data has taken place.

BCD

Binary Coded Decimal: a number representation in which each digit of the number is separately translated into a 4-bit binary code rather than the number being translated completely into pure binary. For example the number 473 would be represented in BCD as 0100 0111 0011.

Binary

1. The number representation system with a base of two (usually using 0 and 1 as digits).
2. A characteristic or property involving a selection, choice or condition in which there are only two possibilities.

Binary Search

A technique for locating data in a file which is in strict sequence, by successively halving the area under search.

Bit

A *bi*nary digi*t*.

Block

A set of data which is of a convenient size to handle as a single unit of transfer between CPU and peripheral device.

Blocking

The combining of two or more records into one block for storage.

Bucket

A term peculiar to direct access storage; a quantity of storage with a fixed position and size dictated by the physical characteristics of the storage device.

Buffer

Temporary storage used to compensate for a difference in the speed of data flow or the occurrence of events when data is being moved from one device to another.

Bureau

An organisation which sells computer time and data processing services (eg, data preparation, programming, etc).

Byte

A contiguous set of binary digits operated upon as a unit (usually comprising eight bits).

Chaining

A method of referencing files, usually randomly stored files, in which each record contains the address, or location, of the next and/or previous records in the file in a predetermined sequence.

Channel

A path linking two hardware or telecommunication units down which data and control signals can be transmitted, eg central processor and a peripheral, or teleprinter terminals; *Multiplexor Channel* – a channel which permits many slow speed input or output devices to communicate simultaneously with the central processor; *Selector Channel* – connects high speed input/output devices to the central processor and operates only one device at a time.

Character

One of the set of symbols that can be used by a particular data processing system, such as the numerals 0 to 9, letters A to Z and additional symbols such as *, +, (,), £, /, –.

Character Rate

The number of characters per unit of time which a device can accept or deliver.

Character Recognition

The machine-reading of characters that are designed to be easily read by human beings; the characters may be in magnetizable ink and read by magnetic ink character recognition (MICR) equipment, or in normal printing ink and read by optical character recognition (OCR) equipment.

Check Digit

A digit (calculated from the digits of an item of data, according to some algorithm) which is appended to the item of data and used to check the validity of the data item subsequently by recalculation.

Check Point

Control point in a program, or file, at intervals to enable re-processing to resume from that point, instead of the start of the program or file should an error occur.

COBOL

A high-level commercial programming language (*Co*mmon *B*usiness *O*riented *L*anguage).

Compaction

Reduction of the size of data items, fields or records by some coding technique.

Compiler

A program which translates a source program written in a statement language into the object, or machine language. Compile, because more than one machine instruction can be generated from a single source instruction. Compare with *assemble*, which is generally a one-for-one, source to object, translation.

Configuration

The complex of interconnected hardware which makes up a computer system.

Copy

To reproduce data in a new location, leaving the data unchanged in its original location.

CPA

Critical Path Analysis: basic technique, using activity networks, for planning and controlling the time schedule of projects. PERT is a development which can incorporate cost aspects and resource allocation.

Critical Path

The sequence of inter-connected events and activities between the start of a project and its completion that will require the longest time to accomplish. This gives the shortest time in which the project can be completed.

Cybernetics

The theory of control and communication in machines and living organisms.

Cycle Time

1. The duration of a complete logical process in respect of a single storage location.
2. The duration of any complete repetitive sequence of operations.
3. Of main store, the minimum time which must elapse between two successive references to a store to insert or extract a word or character.

Cylinder

The tracks on a disk which can be accessed without repositioning the read/write heads (known as a 'seek area').

Database

A collection of data constructed to facilitate the updating once only of the data components and the access and retrieval of individual items. A database is usually designed in such a way as not to restrict its use to a single application.

Database Administrator

The person whose job it is to design, maintain and control the operation of a database.

Data Capture

The process of collecting the data in its various forms for subsequent input into a computer system. Data capture may be either manual or automatic, eg in a badge reading system.

DATAFLOW

A computer assisted method for the recording and analysing of existing or hypothetical systems.

Data Link

A telecommunication system used for transmitting and receiving data between two remote terminals.

Data Preparation

The process of recording raw data in or onto a computer readable medium.

Data Processing

Execution of a systematic sequence of operations upon data, eg merging, sorting, computing.

Data Transmission

Sending data from one place to another, by physical or electronic means.

Debug

To detect, locate, and remove errors from a programming routine or malfunctions from a computer.

Disk

A direct access storage device in which data is recorded on a number of concentric circular tracks on magnetic disks; the required disk and track are selected by electro-mechanical and electronic controls. *Fixed disk:* a disk store which is permanently on-line. *Exchangeable disk:* a disk drive whose store (disk packs) can be removed and stored off-line. *Floppy disk:* a small, flexible disk used mainly for data capture.

Distributed Processing

A data processing system approach which involves centralised design and control but decentralised operation via a number of computers linked in a network.

Drum

A magnetic storage device on which data is stored on a cylindrical drum, subdivided into tracks or bands each having a fixed read/write head. Because the heads are fixed and there is, therefore, no physical movement necessary to access data, the retrieval process is usually very fast. The capacity of drums is, however, small when compared to magnetic tapes and disks.

Dump (v)

To copy stored data onto another, usually more permanent, storage medium.

Edge-punched Card

A card which is punched with holes in a comparatively narrow strip along one edge, allowing printed information to appear on the rest of the card.

Edit

To arrange data into some desired format. For example, to re-arrange and select pertinent data, insert symbols and constants, suppress unwanted zeroes and apply pre-determined format rules to check for validity.

Encode

To translate data into a coded representation. For example to punch data into cards or record onto magnetic tape.

Event

A specified project accomplishment at a particular instant of time (ie no resources are involved) in a project-network.

Exception Report

A report from an information system which highlights exceptions from the normal or anticipated situation. For example, a slow-moving-stock report would list only those items in an inventory which move more slowly than is regarded as acceptable to the designer of the system.

Execute

To perform a data processing routine or program, based on machine language instructions.

Executive (program)

Or, *Supervisor*, *Monitor*, or *Operating System*: a program which controls the use and operation of the equipment and programs.

Field

A portion of or number of associated characters in a peripheral medium, data store or source document used to hold one item of data.

File

An organised collection of usually related information, eg a series of records; *Live File* – a file which is in active use; *Manual File* – a file which is only accessed by people; *File Label* – the identification data, which is recorded on a magnetic storage device, preceding the file to which it refers; *File Mask* – direct – access file protection device which specifies hardware rejection of specific sets of commands.

File Maintenance

The regular processing cycle through which a file goes to ensure that it contains only up-to-date records.

File Organisation

Techniques for storing data on a backing storage device in such a way that the data can be quickly retrieved.

Flag

An additional piece of information added to a data item which gives further information about the data item itself, eg an error flag will indicate that the data item has given rise to an error condition. The term also refers to a single bit or binary data item used to signal whether a state or condition holds or not.

FORTRAN

*For*mula *Tran*slating System. A high-level programming language that closely resembles the algebraic notation of formulae. Primarily for use by engineers and designers.

Gantt Chart

A chart of activity against time used to schedule resources and manpower.

Generation (of file)

When a file of data is updated the new file created is known as a new generation. In processing magnetic tape files it is normal to retain at least three generations as a security precaution against loss or damage to the current generation. These three generations are known as the Grandfather, Father, and Son. Each generation will have the same file name but each will be distinguished from the other by a file generation number which is a number increased by one each time the file is updated.

Graph Plotter

A peripheral device for the production of graphical output from a computer.

Hard Copy

1. A legible copy in conventional characters.
2. A printed copy of machine output, eg printed reports, listings, documents, etc.

Hardware

The physical units from which a computer is built, ie, the mechanical, magnetic, electrical and electronic devices of a computer.

Hash Total

A control total that is the sum of values in a particular field or record area of a file, where the sum has no indicative significance, eg a total of clock numbers.

Heuristic

Serving to discover or invent. Generally used in relation to trial and error method for problem solving.

Hexadecimal

A number system with the base 16. The 16 digit symbols of the system are usually represented by the numbers 0–9 and the letters A–F, (A to F representing the values 10 to 15).

High-level

A classification of computer programming language in which the user is able to write statements in a form oriented towards his usual terminology. The source program must then be subjected to a compilation process to produce the object program in machine code form.

Hit

When processing transactions against a master file only certain members of the file will be matched against an appropriate transaction. When such a match is found this is known as a hit.

Hit Rate

The ratio of records in a master file which are hit in an updating run to the total number of records in the file. If in processing a master debtors' file daily it is found that out of 5,000 records on the file only 250 are updated, the hit rate is 5 per cent.

Home (location)

The location to which a displaced record belongs when overflow takes place.

Information Retrieval

A branch of computing techology related to the storage and categorization of large quantities of information and the automatic retrieval of specific items from the files and indexes maintained.

In-house

Equipment used or activity carried out within a user organisation.

Interactive

A system which allows interaction between man and machine usually via some terminal device.

Interblock Gap

The gap between blocks of data on magnetic tape and disk.

Interrogation

The process of communicating with a file from a terminal.

Interrupt

A break in the normal flow of processing. The normal job flow can be resumed from that point at a later time. An interrupt is usually caused by a signal from an external source, eg a terminal unit, requesting immediate service.

ISDOS

*I*nformation *S*ystem *D*esign and *O*ptimisation *S*ystem – a system developed at the University of Michigan and aimed at automating the systems design process.

Key

An identifier field in a record used for locating or sorting the records.

Key Punch

A machine which incorporates a keyboard used for punching data into punched cards or paper tape.

Kimball Tag

A particular type of perforated card tag attached to merchandise (especially clothing in which the holes represent price, size, colour, style, etc). They are usually converted to punched card or paper tape and the data used for stock control, sales analysis, etc.

Latency

The mechanical delay time in moving a segment of the accessed tape, or disk track under the read/write heads; a component of access time.

Light Pen

A photoelectric light device used in conjunction with a visual display unit by means of which the operator is able to modify the data shown on the screen and the modification is relayed back to the central processor.

Line Printer

A peripheral printing device which prints a line at a time.

List

A series of records each containing a reference, such as an address, to its predecessor and/or its successor.

List Processing

Processing records that are in the form of a list.

Listing

A full print-out of all data in a set of records or file.

Location

An element or group of elements of storage which are uniquely addressable by reference to their location number or address.

Logical

Those aspects of a system or of design concerned with the user's information requirements rather than the physical method used to meet them.

Magnetic Tape

A magnetic storage medium. Magnetic tape is usually ½ in. (12.7 mm) wide and is a plastic tape coated with a magnetic oxide surface. Data is recorded on the tape in frames with each frame holding one character. Magnetic tape is essentially a serial storage medium in that in order to process the data each item on tape must be read from the start of the reel until either the appropriate record is found or the processing of the file is complete.

Magnetic Tape Deck

The transport upon which the reel of tape and its take-up reel are mounted including the read/write/erase head assembly.

Marker

1. A data field, frequently a single bit, used to indicate which of certain possible events has occurred in a program, and subsequently tested to control the path of the program.
2. An item, record or block of data that has a special significance, such as marking the end of a record or file.
3. Signals recorded before and/or after the block of data on a peripheral medium used to indicate the start and/or end of the block.

Mark Reading

The reading of marks on documents or punched cards automatically where the positioning of the mark determines its value. Marks are usually made manually with a pencil. Known as Optical Mark Reading (OMR).

Master File

The file which contains all current data relating to a particular series of records, eg stock master file, debtors' master file.

Medium

Material used in connection with a peripheral device to carry data.

Microfilm

Medium used for recording computer output data, reduced in size and stored on film to be read by a display device.

Multi-access

A system which permits a number of users to access the system at the same time from different points, usually through terminals connected to a communications network. (See also *Time-Sharing*.)

Multiplexing

Simultaneous transmission of a number of different messages over a single circuit; *Multiplexor Channels* – see *Channels*.

Multiprocessor

A machine with multiple arithmetic, logic and main storage units that can be used simultaneously on more than one problem.

Multi-programming

Computer system which permits two or more different application programs to be held (and operative by interleaving) concurrently in main store.

Octal

The octal number system uses a radix or base of 8 so that each position in the number is a power of eight greater than the position immediately to its right. For example, the numbers below are shown in decimal and octal notation. The digits are 0 to 7.

decimal	octal
1 decimal =	1 octal
5	5
8	10
10	12
45	55
473	731

Off-line

Pertaining to those data processing machines or operations that are part of a system but are not directly controlled by the central processor.

On-line

Pertaining to peripheral equipment of the computer, attached to and controlled by the central processor.

Operating System

A complex software system which controls the operation of the computer system and relieves the human operator of much of the detailed work. Normally an operating system will incorporate a complicated scheduling and loading program to optimise utilisation of the available equipment.

Operational Research (OR)

The analytical and mathematical study of human enterprises as an aid to management strategy.

Optical Wand

A hand-held device which reads marks or characters on paper by detecting photo-sensitivity.

Organisation and Methods (O&M)

The systematic review of organisations, their methods and procedures with a view to improving efficiency.

Overflow

The condition arising when the result of an arithmetic/transfer operation is greater than the capacity of the register or storage area in which it is to be stored.

Overlay

A programming technique in which a given area of main store is used for holding successive segments of a program which are too large to be held concurrently.

Overwriting

Updating technique in which the modified record is written back to the original record location in direct access store; the original record is overwritten and therefore lost, unless dumped elsewhere.

Pack (v)

To store several small items in the area normally provided for one word or record.

Package

A major application program written in a general form so that the user is relieved of the effort of programming the bulk of the work and need supply only that data which makes the program specific to his system. Packages now exist in most of the areas where computers are applied, eg payrolls, invoicing, production control.

Packing Density

The amount of information that can be recorded in a given space, for example, the density in bits to the inch on magnetic tape.

Paging

A technique in multiprogramming systems in which programs are divided into blocks or pages of equal size, so that individual pages can be brought into main as required. Normally, special hardware registers are provided to control the indexing of pages. A page is a defined area of backing storage which is transferred into and out of main storage.

Parameter

A quantity or item of information which is used in a routine, subroutine, program, control system or mathematical calculation, and which can be given a different value each time the process is repeated. A subsidiary variable.

Parity Check

An internal computer check to ensure that data is not distorted during a process. Each character is made up of a pattern of 0 and 1 bits plus one extra bit, known as the parity bit. The purpose of the parity bit is to maintain an even or odd (depending on the computer) number of 1 bits. In the event of a bit drop-out the parity check detects that the even or odd ratio is no longer present.

Password

The group of characters input to a computer system by a terminal user in order to gain access to a program or file (if the wrong password is input, access will be prevented).

Peripheral

Auxiliary machine or apparatus under the control of the central processing unit (eg printer).

Physical

Those aspects of a system or of design concerned with the implementation using particular devices, storage methods and programs.

Picture

The method used by COBOL to define the structure of a data item in terms of characters, digits, special symbols, etc.

PL/1

A high-level computer programming language which combines the attributes of COBOL and FORTRAN.

Pointer

A field that contains the address of a storage location that can be used to access that location.

Program

1. The complete sequence of instructions for a job to be performed on a computer.
2. To devise a procedure and list the steps necessary for a job to be performed on a computer. *Object Program* – a program produced by automatic translation of a source program; *Source Program* – a program written in a symbolic language designed for the easy expression of problem-solving procedures; the source program is input to an assembler or compiler and the output is an object program in machine language.

Program Library

The collection of utility programs, subroutines or application programs, used on a particular computer.

Programming Language

Artificial language established for expressing computer programs; *Computer-Oriented Language* (low-level language) – programming language that reflects the structure of a given computer or that of a given class of computers; *Computer Language* (machine language) – computer-oriented language whose instructions consist only of computer instructions: *Assembly Language* – computer-oriented language whose instructions are in one-to-one correspondence with computer instructions with the exception that the language may provide facilities such as macro-instructions; *High-Level Language* – programming language that does not reflect the structure of any one computer or that of any one class of computers.

Radix

The radix, base or basis of a number or notation system is the number of symbols used in the positional representation of the system, ie Binary has a radix of 2, Octal a radix of 8, Decimal a radix of 10. The digit positions in the number system represent the quantity of each power of the radix present in the number, eg $275 = 2 \times 10^2 + 7 \times 10^1 + 5$ (the radix is 10).

Read

To obtain data from one form of store (and transfer it to another form of store).

Real Time

A system designed to produce results in a time-scale which enables them to be used to control the environment producing the input. Examples of real time systems are best found in the process control field where computers are used to control machines.

Response Time

The time required to transmit a message from a terminal to the computer and to receive a reply at that terminal.

Retention Period

The length of time for which a file is to be kept before it may be overwritten. The retention period is written with the file as part of the file label, and, together with the date on which the file was written, is used to calculate the earliest date on which the file may be destroyed. Used as a security precaution against the accidental destruction of a file.

Rotational Delay

The delay experienced on a disk or drum whilst the required portion of the surface is brought under the read/write head.

Search

To examine each item in a set in order to discover whether it satisfies specified conditions.

Seek

1. Synonymous with *Search*.
2. Specialised meaning for electro-mechanical random access devices: to cause the read/write head assembly to move to a given position, eg to a cylinder or set of tracks on a disk store. The track is then *searched* for the desired data record.

Segment

A self-contained part of a program. Frequently programs are divided into segments either because the complete program would be too large to be held in the processor or because a number of programmers are involved and each is to write a different segment.

Sentinel

As marker, a character used to indicate the occurrence of a particular condition.

Sequential Processing

The processing of records in a pre-determined sequence, eg account number or alphabetical order.

Serial Processing

The processing of records strictly in the sequence in which they are stored on the storage medium. Magnetic tape is a serial processing medium whereas magnetic disks may be used for both serial and random processing.

Simultaneity

The simultaneous operation of two or more peripherals or hardware devices under the control of the same program or different programs.

Software

Usually refers to programs written, mainly by a computer manufacturer or software house, and made available to all users of a particular type of computer. Sometimes refers to all the programs which a user has available for use on his machine whether written by his own programmers or outside organisations. The term applies to all sizes and types of program from very simple routines, like transferring data from card to tape, to very complex systems, like production control.

Software House

An organisation which specialises in producing software for users, either purpose-built or generalised.

Spool

An acronym for *S*imultaneous *P*eripheral *O*perations *O*n-*L*ine. Spooling generally means that the computer is occupied with one main job, but for short periods it is able to perform a completely separate and predominantly input-output job (eg printing out a file when the main job is calculating payroll).

Storage

Pertaining to a device in which data can be entered and stored and from which it can be retrieved at a later time.

Switch

A value in a program (which can be changed) used to govern the paths which a program follows.

Synonym

A storage address generated by key transformation which duplicates one already in existence.

Table

An array of data items in list form which is so organised that individual items can be retrieved by specifying keys which are stored as part of the item or by the position of the data item in the table.

TAG

*T*ime *A*utomated *G*rid – a technique developed within IBM to provide computer assistance in systems analysis.

Tagging

The use of short records to represent longer ones in handling sorts or overflow on direct access devices.

Teletypewriter

A terminal device resembling a typewriter used to send and receive messages in a communication system.

Test Data

Data prepared by a systems designer or programmer solely to test the accuracy of the programming and logic of a system. Test data is used to prove each branch of a program and should, therefore, be as comprehensive as possible. Deliberate errors will be introduced into the data, such as inserting alphabetic characters in numeric fields, to ensure that these errors are detected by the program. At the end of the test run the output will be checked against expected results of processing that data to see if they are compatible.

Time Sharing

Carrying out two or more functions during the same time period by allocating small divisions of the total time to each function in turn. To interleave the use of a computer to serve many problem-solvers during the same time span.

Track

The path along which information is recorded on a continuous or rotational medium, such as paper tape, magnetic tape, or a magnetic drum.

Transfer Rate

The rate at which data is moved from a peripheral device to a central processor or vice versa. Usually measured in thousand characters per second (Kch/s).

Truncate

To approximate to the value of a series or number by omitting all digits or terms less than some pre-determined value or after some pre-determined number of terms. In the case of a number, the truncated value is numerically less than the true value.

Turnround Document

One which is produced as output from a system and which, after some intermediate action, is then input, via a reader, back into the system. Examples of these documents are to be seen in public utility billing systems where the computer-produced accounts are encoded in optically readable form and on subsequent payment the bill is used as input to an OCR device. The advantage of such systems is in the elimination of the human element and the consequent danger of introducing errors into the data.

Unpack

To extract several smaller data items from one computer word or record and store them in separate locations.

Update

To bring up-to-date. Usually refers to the operation of incorporating transactions into a master file, such as crediting debtors with cash payments.

User

The person or department using the computer system for the processing of data and the production of information.

Utility

Programs for carrying out regularly required routine operations such as copying data from punched cards to tape or disk, sorting files into sequence.

Validity Check

A check carried out on data, usually on initial input to the system, to ensure as far as possible that it is valid. Such checks include looking for data which exceeds stated limits, or checking that alphabetic fields do not include numeric characters and vice versa.

Variable Length

A method of storing data where each record is not necessarily the same number of characters in length as any other record. The end of a record is identified by a special end-of-record character.

Verification

The process of checking the accuracy of punching on paper tape and punched cards.

Virtual Memory

A store management system in which a user uses the storage resources of the computer without being constrained by the limited size of mainstore.

Visual Display Unit (VDU)

An input/output device equipped with a television type screen, eg a cathode ray tube, and a keyboard thus permitting data to be displayed on a screen and providing facilities for the operator to input data to the system via the keyboard.

Word

A basic unit of data in a computer memory; a word consists of a predetermined number of characters, bits or bytes to be processed as an entity, eg a program instruction or an element of data. A word may be the smallest addressable element of a store or may be further subdivided into bytes, each separately addressable. In many digital computers a fixed length word is used, but in other machines characters may be grouped to form words of variable length according to the requirements of the particular instruction to be performed.

Work Study (WS)

The systematic examination of work procedures (normally on the shop floor) with a view to improving efficiency.

Write

To copy or transcribe (ie altering the form or representation of) data into one form of store from another.

Write Ring

A device attached by the operator to a spool of magnetic tape when writing is to take place; its absence prevents writing.

Appendix E
Index